Stochastic Geometry Analysis of Multi-Antenna Wireless Networks

Xianghao Yu · Chang Li ·
Jun Zhang · Khaled B. Letaief

Stochastic Geometry Analysis of Multi-Antenna Wireless Networks

Xianghao Yu
Department of Electronic
and Computer Engineering
Hong Kong University of Science
and Technology
Hong Kong, China

Chang Li
Department of Electronic
and Computer Engineering
Hong Kong University of Science
and Technology
Hong Kong, China

Jun Zhang
Department of Electronic
and Information Engineering
Hong Kong Polytechnic University
Kowloon, Hong Kong, China

Khaled B. Letaief
Department of Electronic
and Computer Engineering
Hong Kong University of Science
and Technology
Hong Kong, China

ISBN 978-981-13-5879-1 ISBN 978-981-13-5880-7 (eBook)
https://doi.org/10.1007/978-981-13-5880-7

Library of Congress Control Number: 2019934527

This Springer imprint is published by the registered company Springer Nature Singapore Pte Ltd.
The registered company address is: 152 Beach Road, #21-01/04 Gateway East, Singapore 189721, Singapore

To my beloved parents

Xianghao Yu

Dear Shuang, will you marry me?

Chang Li

To Lori, Ollie, and my parents

Jun Zhang

To Selma, Amina, Leyla, and Elissa

Khaled B. Letaief

Preface

The past decades have witnessed the revolution brought by wireless communications technologies, which have fundamentally affected major economic sectors. In particular, mobile Internet has enabled various innovative applications, notably mobile payment, mobile e-commerce, and abundant online-to-offline (O2O) applications. This trend in turn has led to the growing popularity of smart mobile devices. A consequence is that global mobile data traffic has been growing continuously at an exponential rate. The capacity of wireless networks thus has to increase substantially to meet the explosive demands. For example, the upcoming 5G networks aim at carrying out the projected 1000X increase in network capacity by 2020, in comparison with current 4G LTE networks. One way to boost the capacity is to improve the spectral efficiency via deploying multiple antennas at access points, such as massive multiple-input multiple-output (MIMO) systems. Further improvement in area spectral efficiency can be achieved by network densification, i.e., deploying more and more radio access points. These are two dominant trends in future wireless network evolution, which motivate the main theme of this monograph, namely, the performance analysis of large-scale multi-antenna wireless networks. Chapter 1 provides more background discussions.

Stochastic geometry, in particular point process theory, is adopted as the main analytical tool. Chapter 2 introduces some preliminaries of stochastic geometry. While being rigorous, this chapter is not pedantic and does not dwell on the detailed theory of point processes. The interested reader can always study these intricacies from mathematical literatures. To illustrate the principle of network performance analysis via stochastic geometry, existing approaches and results for single-antenna networks are presented, which also serve as a basis for the later chapters.

After introducing typical multi-antenna transmission techniques, a general analytical framework is presented for multi-antenna wireless networks in Chap. 3. Detailed derivations are provided to the readers who are interested in the intrinsic mechanism of the framework; other readers may directly visit the well-rounded results, e.g., Theorem 3.2, Proposition 3.1, and Proposition 3.2, and apply them as a toolbox to the network of interests. The presented framework entails the existing approaches for single-antenna networks as special cases and leads to both tractable and insightful analytical results for multi-antenna networks.

To illustrate the effectiveness of the presented framework, four application examples are given in Chaps. 4 and 5. Readers may get a clearer idea of how they can apply this framework by following these specific examples with different network models. In particular, a variety of networks are investigated with tractable analytical results, including multiple-input single-output (MISO) small cell networks, millimeter-wave networks, and MIMO heterogeneous networks. More importantly, abundant network design insights and guidelines are unraveled by the framework, e.g., the impacts of the BS density and antenna size on the coverage probability, area spectral efficiency, and energy efficiency.

The monograph is suitable for both beginners in wireless network analysis and experienced researchers who intend to work in depth on the performance analysis of multi-antenna wireless networks. In the former case, some parts in Chap. 2 should be carefully followed, and the references therein may be referred for further explanation. Readers who are more interested in specific wireless network models may pay more attention to details in Chaps. 4 and 5. Furthermore, Chap. 6 provides some potential future research directions for further study and investigation.

Wireless network analysis via stochastic geometry has received significant attention in recent years. There have been a few textbooks and monographs on the subject, and this one differentiates by focusing on multi-antenna wireless networks. The main motivations are twofold. On the one hand, most modern wireless communications systems have adopted multi-antenna transmissions, and thus their performance analysis is of significant practical importance. On the other hand, existing references mainly focused on single-antenna network analysis, which cannot be directly applied to the multi-antenna case. We intend to fill this gap and present in this monograph a set of analytical methodologies for large-scale multi-antenna wireless networks. While the network models treated in this monograph have certain limitations, we hope the presented results will inspire more researchers to work on this important and exciting area.

Acknowledgements

The works in this monograph have been supported by the Hong Kong Research Grants Council No. 610113 and No. 16210216. We also would like to express our sincere thanks to Dr. Martin Haenggi and Dr. Jeffrey G. Andrews for their valuable contributions to the related works presented in this monograph.

Erlangen, Gaithersburg, Hong Kong
January 2019

Xianghao Yu
Chang Li
Jun Zhang
Khaled B. Letaief

Contents

Acronyms

1G	The first generation
2G	The second generation
3G	The third generation
4G	The fourth generation
5G	The fifth generation
6G	The sixth generation
ADC	Analog-to-digital converter
AoA	Angle of arrival
AoD	Angle of departure
AP	Access point
AR	Augmented reality
ASE	Area spectral efficiency
BD	Block diagonalization
BPP	Binomial point process
BS	Base station
ccdf	Complementary cumulative distribution function
CCI	Co-channel interference
CDI	Channel direction information
CSI	Channel state information
D2D	Device-to-device
DPC	Dirty paper coding
eMBB	Enhanced mobile broadband
FDD	Frequency division duplexing
GB	Gigabyte
GPS	Global Positioning System
HetNet	Heterogeneous network
HPBW	Half-power beamwidth
IN	Interference nulling
IoE	Internet of everything
IoT	Internet of things

LOS	Line-of-sight
LTE	Long-term evolution
MEC	Mobile edge computing
MIMO	Multiple-input multiple-output
mm-wave	Millimeter-wave
mMTC	Massive machine-type communications
MRT	Maximum ratio transmit
MU-MIMO	Multiuser MIMO
NLOS	Non-line-of-sight
pdf	Probability density function
PGFL	Probability generating functional
PHP	Poisson hole process
PPP	Poisson point process
QoE	Quality of experience
RVQ	Random vector quantization
SINR	Signal-to-interference-plus-noise ratio
SIR	Signal-to-interference ratio
SVD	Singular vector decomposition
TC	Transmission capacity
TDD	Time division duplexing
TDMA	Time division multiple access
THP	Tomlinson–Harashima precoding
UAV	Unmanned aerial vehicle
ULA	Uniform linear array
URLLC	Ultrareliable low-latency communications
UR-SP	Uniformly random single path
V2I	Vehicle-to-infrastructure
V2V	Vehicle-to-vehicle
ZF	Zero-forcing
ZF-RAS	Zero-forcing with receive antenna selection
ZF-TAS	Zero-forcing with transmit antenna selection

Chapter 1
Introduction

Abstract This chapter starts with the latest trends in wireless network evolution and introduces different modeling and analysis techniques for network performance evaluation. The evolution of cellular networks is first presented. Then three unique characteristics of the latest cellular standard, i.e., the fifth generation (5G) network, are illustrated. Through the discussion, it is revealed that network densification and multi-antenna transmissions are two main enabling techniques to achieve the targets of 5G, which motivates the main theme of this monograph, i.e., performance analysis of large-scale multi-antenna wireless networks. Next, an overview of network modeling and analysis is introduced, followed by the outline of the monograph.

1.1 The Cellular Network Evolution

During the past decades, we have witnessed the phenomenal development of wireless communications, with expanding social and economic impacts. The first generation (1G) mobile communications system was introduced in the 1980s, where the radio signal was analog and only basic voice services were provided. In the 1990s, the second generation (2G) system was developed, where the digital communication techniques were introduced to wireless communications systems. Compared with 1G systems, 2G technologies enhanced the voice service quality and enabled services such as text messages. Later on, the third generation (3G) system was introduced in the 2000s with the peak data rate up to 2 Mbps, which supports Global Positioning System (GPS)-based services, mobile video calls and TV streaming, etc. [1]. To integrate 3G with the fixed Internet to support wireless mobile Internet, the fourth generation (4G) mobile communications system was deployed since 2009 based on the Long-Term Evolution (LTE) standard [2], which provides high-speed wireless communications for more sophisticated mobile devices. Thanks to the mobile Internet, many innovative applications have been proposed and realized, such as amended mobile web access, mobile gaming services, mobile payment, high-definition mobile TV, and video conferencing. More recently, driven by the increasing demands, the fifth generation (5G) cellular standard has been developed, aiming at achieving the transmission data rate as tens of gigabytes per second. Particularly, 5G

X. Yu et al., *Stochastic Geometry Analysis of Multi-Antenna Wireless Networks*, https://doi.org/10.1007/978-981-13-5880-7_1

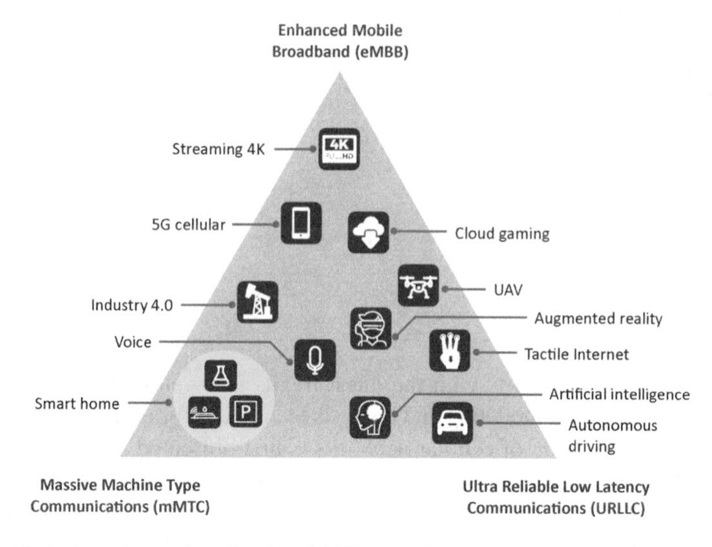

Fig. 1.1 Typical services and applications in 5G networks

is not merely an increment of the previous 4G system. Instead, it is a paradigm shift that meets unique requirements in modern wireless networks [3]. Specifically, the 5G requirements for IMT-2020 include the support of various services and applications, which can be grouped into three categories, namely enhanced mobile broadband (eMBB) communications, massive machine-type communications (mMTC), and ultra-reliable low-latency communications (URLLC). Several typical applications for these scenarios are exemplified in Fig. 1.1, and the specifics are illustrated below.

According to CISCO, by 2017, each user owned on average 2.3 devices, the data traffic for each user per month reached 24 Gigabyte (GB), and 67% of them was video traffic. It is predicted that these three numbers will keep increasing, and by 2021, there will be 3.5 devices/user, 61 GB data/user/month, and 80% of them will be multimedia traffic that needs ultra-high-data rate transmission [4]. All these facts require the newly proposed 5G system to enhance the capacity for mobile broadband communications. In particular, the upcoming 5G networks aim at carrying out the projected 1000X increase in capacity by 2020 to achieve eMBB communications [5, 6].

Meanwhile, there are many emerging concepts related to 5G systems, e.g., the Internet of Things (IoT), Internet of Everything (IoE), Industry 4.0, and Smart X. These application scenarios demand effective connection between millions of sensors, machines, factories, and infrastructures, which is defined as the mMTC. The main challenge in mMTC is to provide scalable and efficient connectivity for a massive number of devices sending very short packets. This cannot be accommodated adequately in current cellular systems that are mainly designed for human-type communications.

Besides the requirement of massive connectivity, 5G networks also target at low-latency communications [7]. Round-trip latencies in current 4G networks are on the order of about 15 ms. However, with new applications emerging, such as augmented reality (AR) and the tactile Internet, the latency in 5G systems should be reduced to within 1 ms [8]. In addition, the reliability of the current wireless systems needs to be further improved as some 5G applications are extremely sensitive to communications uncertainty, such as remote surgical consultations and surgery. In this way, new communications paradigms, such as mobile edge computing (MEC), edge caching at radio access points (APs), and device-to-device (D2D) communications, are required to help realize URLLC communications.

In summary, the 5G wireless network is expected to provide significant network capacity increase, support massive connectivity, and achieve ultra-low transmission latency. Furthermore, the evolution of wireless networks never stops, and beyond 5G systems, i.e., the sixth generation (6G) and beyond, have already drawn much academic and industrial attention nowadays while the 5G system is still on the way of real implementation. In this monograph, we shall focus on network analysis of large-scale wireless networks, including 5G and beyond. While the major examples are from 5G networks, the methodologies introduced are more broadly applicable.

1.2 Paving the Way for 5G Networks

Among the three main application scenarios for 5G, we shall pay special attention to eMBB. This aspect has drawn the most attention from both the academia and industry. Meanwhile, the techniques developed for eMBB will also play vital roles in achieving mMTC and URLLC. In this section, we introduce three key technologies, as illustrated in Fig. 1.2, to enhance the capacity of 5G networks, and thus achieve the target of eMBB.

1.2.1 Ultra-Dense Networks to Increase Spatial Spectrum Reuse

Network densification, i.e., deploying more APs, has been proven to be an effective way to improve the network capacity [9, 10]. This, in turn, makes the coverage of each AP smaller than the conventional macro base stations (BSs), which is called a *small cell* AP [11]. These APs typically operate with lower power and cost, which meanwhile improves the energy efficiency of the network. By densely deploying different types of small cells overlaid with the existing macrocell network, we essentially obtain a heterogeneous network. In this way, spatial spectrum reuse and thus area spectral efficiency (ASE) can be significantly improved, and uniform coverage can be provided.

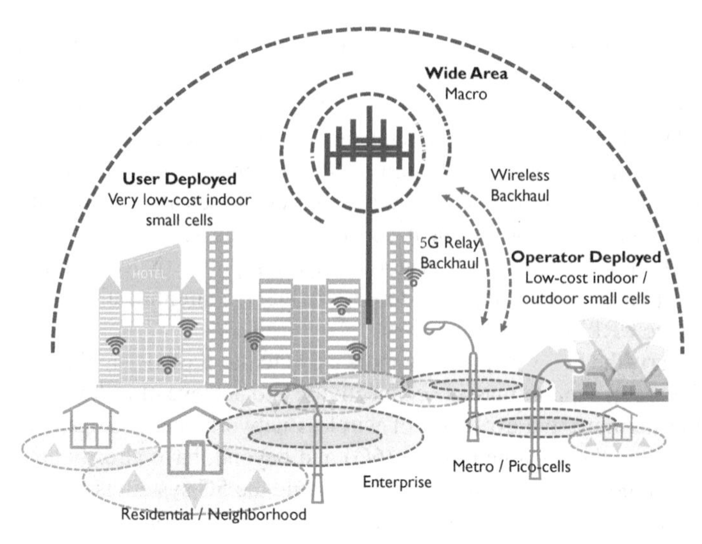

(a) Ultra-dense networks to increase spatial spectrum reuse.

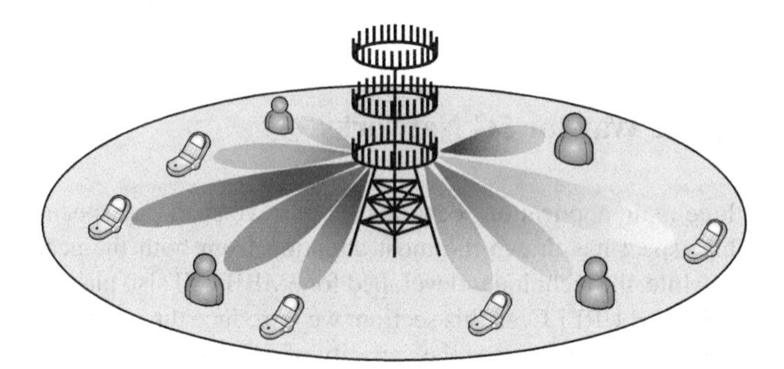

(b) Multi-antenna transmissions for higher spectral efficiency.

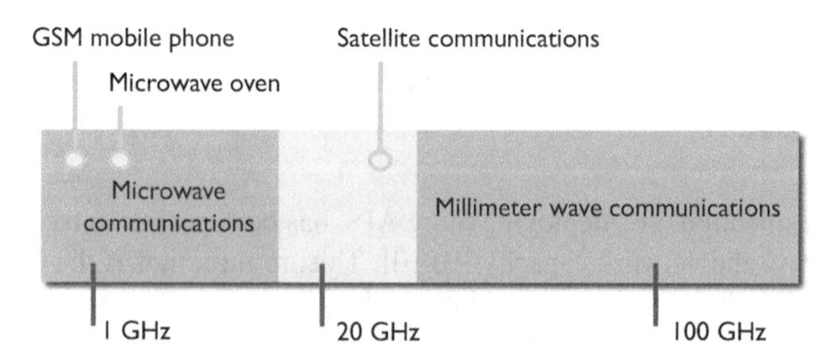

(c) Beyond sub-6 GHz for more spectrum.

Fig. 1.2 Three key technologies to enhance the capacity of 5G networks

A holistic view on ultra-dense small cell and heterogeneous networks (HetNets) was given in [12]. In particular, several technical challenges have been identified, i.e., intercell interference, unplanned deployment, mobility management, and privacy in small cell networks. As the first step for designing ultra-dense wireless networks, characterizing the network performance is needed to quantify the performance gain of network densification and to help reveal network design insights. There have been many existing works on performance analysis of small cell networks. A well-known result [13] for small cell networks showed that, with certain assumptions on the network and channel models, the downlink signal-to-interference ratio (SIR) coverage is invariant with the network densification. This indicates that deploying more APs will definitely benefit the network throughput without deteriorating the received SIR. In addition, the authors of [14] derived an accurate model for HetNets, based on which performance analysis in terms of the coverage probability was carried out. One interesting observation for interference-limited open access HetNets is that, similar to small cell networks, adding more tiers and/or APs in HetNets neither increases nor decreases the probability of coverage or outage when all the tiers have the same target SIR. Thus, the same as the single-tier network, the network capacity of HetNets increases linearly with the density of APs. This line of results also illustrates the powerfulness of analytical network performance analysis, which will be shown in full details later.

1.2.2 Multi-Antenna Transmissions for Higher Spectral Efficiency

Network densification aims at increasing the number of cells per unit area to improve the network capacity. On the other hand, increasing the spectral efficiency in each cell, i.e., the number of transmitted bits per second per Hertz, is another approach to enhance the network capacity. Multiple-input multiple-output (MIMO) technique, i.e., multi-antenna transmissions, is one of the main techniques for achieving high spectral efficiency, which leverages a large number of antennas to provide significant beamforming gains to combat the path loss and suppress interference. There are two flavors of MIMO with respect to how data is transmitted across the wireless channel. Aiming at improving the transmission reliability, we may choose to send the same data via different spatial propagation paths, which reduces error probability via *spatial diversity*. On the other hand, multiple data streams can be simultaneously transmitted on different propagation paths for improving the data rate of the system, which is called *spatial multiplexing* [15].

Various MIMO transmission techniques have been employed in broadband wireless systems, such as WiFi and 4G LTE networks. Recently, massive MIMO techniques have been proposed for 5G systems, where hundreds of antennas can be leveraged, and extensive ongoing research has been done in this area. Many benefits of massive MIMO systems have been identified [16, 17]. For example, massive

MIMO can be built with inexpensive and low-power components. It enables a significant reduction of latency on the air interface, simplifies the multiple access layer, and increases the robustness against both unintended man-made interference and intentional jamming. Furthermore, with more antenna elements compared with conventional MIMO systems, various beamforming techniques can be leveraged for improving the directionality of the transmit and receive beams [18]. Finally, spectral efficiency can be further improved by combining the multiuser transmission and massive MIMO techniques, which offers big advantages over the conventional point-to-point MIMO [16]. In particular, multiuser MIMO can work with low-cost single-antenna terminals, in relatively poor scattering environment, and with simplified resource allocation. The performance of massive MIMO systems depends critically on the availability of channel state information (CSI). Ideally, every terminal in a massive MIMO system is assigned with an orthogonal uplink pilot sequence for CSI acquisition. However, the maximum number of orthogonal pilot sequences is limited by the available radio resource, and the same pilot sequence may be used in multiple cells. Therefore, pilot contamination caused by pilot sequence reuse is a crucial problem to tackle [19].

Massive MIMO systems can be deployed with both the time division duplexing (TDD) and frequency division duplexing (FDD) modes. The original massive MIMO concept [20, 21] assumes TDD, where the uplink and downlink transmissions take place in the same spectrum band but are separated sequentially in the time domain. On the other hand, motivated by spectrum regulation issues, there is significant interest in developing FDD versions of massive MIMO [22], where separate frequency bands are used for simultaneous transmission at the transmitter and receiver sides. TDD systems have the advantage in CSI acquisition by exploiting the reciprocity of the uplink and downlink channels within the coherence time in the same band. However, one piratical problem for implementing TDD massive MIMO is that different operators might need to synchronize their base stations, to avoid that one operates in the downlink and one in the uplink on adjacent frequency bands. Such a problem can be easily avoided in FDD by separating the uplink and downlink bands, which makes it a more commonly adopted protocol nowadays. It is expected that both TDD and FDD massive MIMO will be deployed in 5G networks.

1.2.3 Beyond Sub-6 GHz for More Spectrum

In the previous two approaches, the number of nodes per unit area and the spectral efficiency per node are increased to improve the network capacity. On the other hand, it is expected that more bandwidth will be needed in 5G systems to further improve capacity. Currently, the spectrum in use for communications systems is typically below 6 GHz, referred to as the *sub-6 GHz spectrum* [23]. However, the spectrum crunch in current cellular systems brings a fundamental bottleneck for the further capacity increase. Thus, it is critical to exploit underutilized spectrum bands, including the bands that have not been used for cellular communications yet. Millimeter-wave (mm-wave) bands from 30 GHz to 300 GHz, previously only considered for outdoor point-to-point backhaul links [24] or for carrying indoor high-resolution multimedia streams [25], have now been put forward as a prime candidate for new spectrum in 5G cellular systems, with the potential bandwidth reaching 10 GHz. This view is supported by recent experiments in New York City that demonstrated the feasibility of mm-wave outdoor cellular communications [26].

The main obstacles for the success of mm-wave cellular systems include the huge path loss and rain attenuation, as a result of the tenfold increase of the carrier frequency [26]. Thanks to the smaller wavelength of mm-wave signals compared with sub-6 GHz systems, more antenna elements can be patched together in a small area, and therefore large-scale antenna arrays can be leveraged at transceivers to provide significant beamforming gains to combat the path loss and to synthesize highly directional beams. In this way, the use of mm-wave bands is naturally compatible with the massive MIMO evolution for 5G systems. Another obstacle for mm-wave communications is hardware implementation. With large-scale antenna arrays, the number of hardware components at transceivers increases dramatically. More importantly, the hardware components for communications at mm-wave frequencies are difficult to develop, such as the high-speed high-accuracy analog-to-digital converters (ADCs) and signal mixers [23]. In addition, mm-wave systems will suffer huge noise power due to the increased bandwidth. These challenges have triggered many ongoing research efforts, for example, in cost-effective beamforming [27, 28], initial access [29], and channel estimation [30].

1.2.4 *Two Dominant Themes*

The above discussions on the three key approaches for capacity increase in 5G networks reveal the critical roles of network densification and multi-antenna transmissions. On the one hand, the dense small cells significantly increase the spatial reuse of the limited spectrum, and also make the communication distance shorter than that in conventional systems. The short-distance communications facilitates the deployment of mm-wave communications whose propagation range is relatively short due to high path loss. On the other hand, the massive MIMO technique validates the use of a huge number of antennas elements that are patched together in a small area at mm-wave frequencies, and can effectively exploit them to improve spectral efficiency and energy efficiency. In addition, multi-antenna nodes have better capability to suppress intercell interference, which is the limiting factor in dense networks. Thus, the analysis, understanding, and design of large-scale multi-antenna networks are critical for 5G networks, which form the main subject of this monograph.

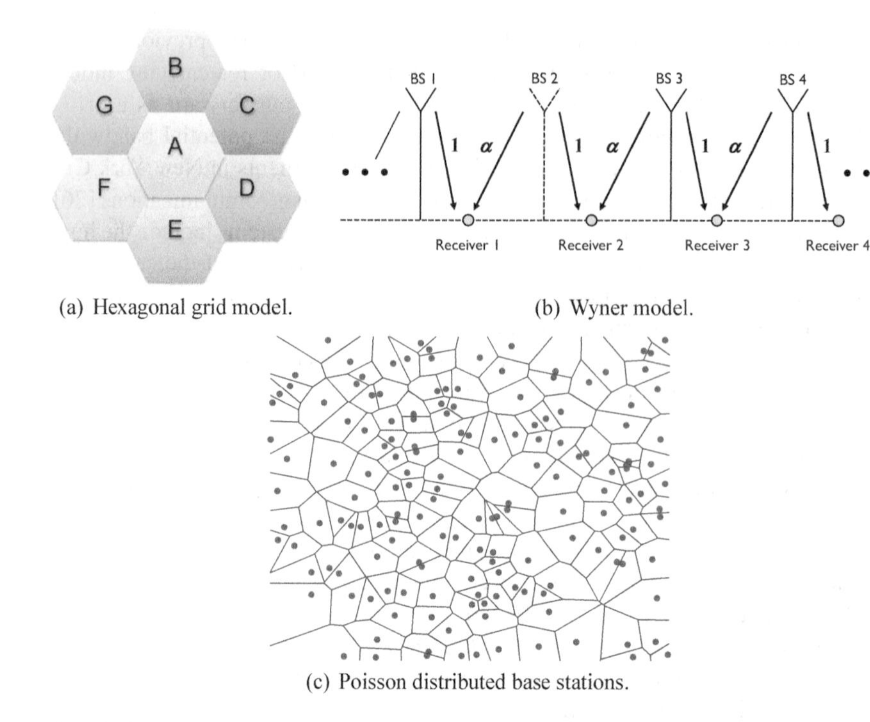

(a) Hexagonal grid model.

(b) Wyner model.

(c) Poisson distributed base stations.

Fig. 1.3 Comparisons between the grid model, Wyner model, and PPP model

1.3 Modeling and Analysis of Wireless Networks

It is crucial and intriguing to investigate and characterize the performance of wireless networks. As field trials are costly and system-level simulations are time-consuming, an effective alternative is to evaluate the network performance analytically with the assistance of a mathematical model. With the analytical results of the wireless networks at hands, we are then able to reveal network design insights by investigating the impact of different network parameters and to optimize the network for different performance metrics, e.g., spectral efficiency, energy efficiency, and loading fairness. Thus, performance analysis of wireless networks is not only of theoretical interest but also of significant practical importance.

To analytically characterize the performance of wireless networks, a tenable and tractable mathematical tool is a basis. The network performance is influenced by the locations of the APs and users, and therefore an accurate mathematical model for the location distributions of these nodes is required to help depict the network performance. Cellular networks were traditionally modeled by placing the base stations on a grid, with mobile users either randomly scattered or placed deterministically. A simplified and typically adopted way is to model the spatial distribution of cellular networks as an equilateral triangular lattice, which essentially is a *hexagonal grid model*. In the example shown in Fig. 1.3a, there are seven sites and each of them covers a geographical zone, called a Voronoi cell. In the early days of analyzing cellular networks, this model was widely used to characterize the co-channel interference, and determine the frequency reuse pattern [31]. A recent study on tractable performance analysis based on the hexagonal cellular network model can be found in [32]. Generally speaking, performance analysis with the grid model relies on lots of simplifications and approximations, and exact analysis typically does not lead to tractable results. Moreover, cell shapes in modern wireless networks become highly irregular, and thus the accuracy of the grid model further deteriorates. A more tractable model is the Wyner model [33], which has been commonly adopted in the information theoretic studies of the capacity of cellular networks [34]. Its key aspects include fixed user locations and the deterministic and homogeneous interference intensity. In particular, only interference from two adjacent cells is considered for each receiver, as shown in Fig. 1.3b. Such simplification helps to derive nice closed-form expressions for the uplink and downlink capacity, but the accuracy is questionable. It was shown in [13] that the grid model provides an upper bound to the spatial distribution in reality, whereas the Wyner model is an oversimplified and highly inaccurate model unless there is a very large amount of interference averaging over space, such as in the uplink of heavily loaded CDMA systems [35, 36].

To better model the irregularity of the spatial distribution of modern wireless networks, random spatial network models were then proposed for network performance analysis. The transport capacity was introduced and first quantified in [37] by assuming that a finite number of nodes in an ad hoc network are either arbitrarily located or uniformly distributed. Furthermore, upper and lower bounds on the asymptotically achievable throughput were also derived. Based on the random spatial models, the effects of mobility on the transport capacity of ad hoc networks were analytically characterized in [38]. It was shown that the per-user throughput can increase dramatically when nodes are mobile rather than fixed, which can be achieved by exploiting a form of multiuser diversity via packet relaying. Nevertheless, these results are less specific in the sense that only asymptotic results or bounds are available, mainly due to the lack of tractability of the adopted random spatial models.

Stochastic geometry is a rigorous mathematical tool for studying problems with spatially randomly distributed quantities, and has found a number of applications, e.g., the statistical theory of shape, in material science, and image analysis [39]. Stochastic geometry, especially point process theory, has recently been applied to wireless networks to model the spatial distribution of the transceivers, and the past decade has witnessed a significant growth in this area [40]. In particular, a special point process, called the *Poisson point process (PPP)*, has been proposed to model the spatial distribution of the transceivers in wireless networks. It has been shown that the PPP model is especially accurate at describing the unique characteristics in dense networks [13, 41], by comparing Fig. 1.3c with [13, Fig. 2]. A well-known example of analyzing wireless networks based on the PPP model is [42], where the transmission capacity [43] was derived in closed forms.

For cellular networks, adopting the PPP model for the analysis was first advocated in [41], which derived tractable results for the coverage and ergodic rate analysis. It disclosed that the outage probability is critically determined by the Laplace transform of the aggregate interference. While this study inspired many research works on network analysis and design, e.g., [44, 45], the main approach is only applicable to certain network settings, including the Rayleigh fading channel, single-slope path loss, and, more importantly, single-antenna transceivers. As illustrated above, deploying multiple antennas in dense networks is one of the prevalent dimensions for future network evolution. Unfortunately, extending the existing analytical approach to more advanced network settings, e.g., multi-antenna networks, is highly nontrivial. In particular, with various multi-antenna transmission techniques, the distributions of both the desired signal and aggregated interference are much more complicated than those in conventional single-antenna systems, with which the network performance does not simply depend on the Laplace transform itself. This calls for a novel and systematic approach for performance analysis of networks that incorporate advanced MIMO techniques. Recently, we have witnessed a series of works on deriving tractable analytical results for multi-antenna wireless networks, leading to insightful observations and effective designs for various network models [46–50]. Introducing this analytical framework is the main subject and shall be discussed in detail in this monograph.

1.4 Outline of the Monograph

There exist several textbooks and monographs on the topic of wireless network analysis via stochastic geometry [40, 51, 52]. General introduction of stochastic geometry and its applications in wireless communications were presented in [40]. The analysis of transmission capacity in ad hoc networks via stochastic geometry was presented in [51]. Nevertheless, these two books only considered single-antenna wireless networks. Multi-antenna network performance analysis via stochastic geometry was carried out in [52]. However, the analytical results therein are either asymptotic ones for massive MIMO systems based on the random matrix theory, or approximate ones. In contrast, this monograph presents a systematic and accurate approach to analyze the performance of general multi-antenna wireless networks, as well as its applications for network optimization. Following is an outline of the monograph.

In Chap. 2, preliminary concepts and useful analytical techniques of stochastic geometry are introduced. In particular, several commonly encountered point processes are first presented, including the binomial point process and Poisson point process. Next, key performance metrics of wireless networks are defined, such as the coverage probability, outage probability, spectral efficiency, energy efficiency, and area spectral efficiency. One critical analytical task, i.e., characterizing the aggregated interference, is then illustrated in detail. Typical analytical results in single-antenna wireless networks are also provided. This chapter forms the basis of the investigation in the following chapters.

In Chap. 3, using tools from stochastic geometry, a unified framework is presented for analyzing large-scale multi-antenna wireless networks. The major results are two innovative representations of the coverage probability, which make the analysis of multi-antenna networks almost as tractable as the single-antenna case. One is expressed as an ℓ_1-induced norm of a Toeplitz matrix, and the other is given in a finite sum form. With a compact representation, the former incorporates many existing analytical results on single- and multi-antenna networks as special cases, and leads to tractable expressions for evaluating the coverage probability in both ad hoc and cellular networks. While the latter is more complicated for numerical evaluation, it helps analytically gain key design insights. In particular, it helps prove that the coverage probability of ad hoc networks is a monotonically decreasing convex function of the transmitter density and that there exists a peak value of the coverage improvement when increasing the number of transmit antennas. On the other hand, in multi-antenna cellular networks, it is shown that the coverage probability is independent of the transmitter density and that the outage probability decreases exponentially as the number of transmit antennas increases.

The general analytical framework is then applied to different types of networks in Chap. 4 to reveal various network insights analytically. First of all, we investigate the effect of network densification in multi-antenna small cell networks. It is analytically shown that deploying more BSs always increases the network throughput, but the throughput will scale with the BS density first linearly, then logarithmically, and finally converge to a constant. The energy efficiency of multi-antenna small cell networks is then analyzed. Specifically, increasing the BS density or the number of transmit antennas will first increase and then decrease the energy efficiency if the non-transmission power or the circuit power consumption is less than certain thresholds, and the optimal BS density and the optimal number of BS antennas can be found. Otherwise, the energy efficiency will always decrease.

Furthermore, as one of the evolution paths for future networks, the performance of mm-wave networks is characterized. Directional transmission is one unique feature in mm-wave networks compared with sub-6 GHz ones. To help yield insights on the role of directional antenna arrays in mm-wave networks, we present two sophisticated antenna patterns with desirable accuracy and analytical tractability. We have shown that the coverage probabilities of both ad hoc and cellular networks increase as a non-decreasing concave function with the antenna array size. In addition, asymptotic outage probabilities are also derived when the number of antennas approaches infinity, which are shown to be inversely proportional to the array size.

In Chap. 5, based on the general analytical framework, methodologies for network optimization are presented. First, a novel interference coordination strategy for multi-antenna cellular networks, called user-centric intercell interference nulling (IN), is introduced for dense small cell networks. The main merit of this strategy is its ability to effectively identify and mitigate the dominant interference for each user. We derive an approximate expression of the coverage probability of the multi-antenna network with this coordination strategy, based on which the optimal IN range is analytically determined.

Finally, we analyze and optimize more general wireless networks, i.e., multi-antenna HetNets. The main analytical results reveal that the SIR invariance property in single-antenna HetNets does not hold for MIMO HetNets; instead, the success probability may decrease as the network density increases. We prove that the maximum success probability is achieved by activating only one tier of BSs, while the maximum ASE is achieved by activating all the BSs. This reveals a unique trade-off between the ASE and link reliability in multiuser MIMO HetNets. To achieve the maximum ASE while guaranteeing a certain link reliability, we develop efficient algorithms to find the optimal BS density for each tier of the network.

Last but not least, Chap. 6 summarizes the monograph, along with discussions on promising future research directions. In particular, the analytical results presented in this monograph are first summarized. Then, potential extensions of the presented analytical framework are discussed. The framework can be extended to more generic network models by generalizing the path loss model, small-scale fading, MIMO transmission strategies, cell association strategies, and adopted point processes. On the other hand, applying the analytical framework to newly emerged applications is also interesting to explore. Extensions to unmanned aerial vehicle systems, physical layer security-aware networks, and vehicular communications systems are discussed.

References

1. C. Smith, *3G wireless networks* (McGraw-Hill, Inc., 2006)
2. A. Ghosh, J. Zhang, J.G. Andrews, R. Muhamed, *Fundamentals of LTE* (Pearson Education, 2010)
3. J.G. Andrews, S. Buzzi, W. Choi, S.V. Hanly, A. Lozano, A.C.K. Soong, J.C. Zhang, What will 5G be? IEEE J. Sel. Areas Commun. **32**(6), 1065–1082 (2014)
4. Cisco visual networking index: Forecast and methodology, 2016–2021 (2017)
5. Q.C. Li, H. Niu, A.T. Papathanassiou, G. Wu, 5G network capacity: key elements and technologies. IEEE Veh. Techn. Mag. **9**, 71–78 (2014)
6. C. Wang, F. Haider, X. Gao, X. You, Y. Yang, D. Yuan, H.M. Aggoune, H. Haas, S. Fletcher, E. Hepsaydir, Cellular architecture and key technologies for 5g wireless communication networks. IEEE Commun. Mag. **52**, 122–130 (2014)
7. R. Vannithamby, S. Talwar, *Low-latency radio-interface perspectives for Smallcell 5G networks* (Wiley, 2017)
8. J. Pilz, M. Mehlhose, T. Wirth, D. Wieruch, B. Holfeld, T. Haustein, A tactile internet demonstration: 1 ms ultra low delay for wireless communications towards 5G, in *IEEE Conference Computer Communications Workshops*, pp. 862–863, Apr. 2016
9. N. Bhushan, J. Li, D. Malladi, R. Gilmore, D. Brenner, A. Damnjanovic, R.T. Sukhavasi, C. Patel, S. Geirhofer, Network densification: the dominant theme for wireless evolution into 5G. IEEE Commun. Mag. **52**, 82–89 (2014)
10. X. Ge, S. Tu, G. Mao, C. Wang, T. Han, 5G ultra-dense cellular networks. IEEE Wirel. Commun. **23**, 72–79 (2016)
11. J. Hoadley, P. Maveddat, Enabling small cell deployment with HetNet. IEEE Wirel. Commun. **19**, 4–5 (2012)
12. I. Hwang, B. Song, S.S. Soliman, A holistic view on hyper-dense heterogeneous and small cell networks. IEEE Commun. Mag. **51**, 20–27 (2013)
13. J.G. Andrews, F. Baccelli, R.K. Ganti, A tractable approach to coverage and rate in cellular networks. IEEE Trans. Commun. **59**, 3122–3134 (2011)
14. H.S. Dhillon, R.K. Ganti, F. Baccelli, J.G. Andrews, Modeling and analysis of K-tier downlink heterogeneous cellular networks. IEEE J. Sel. Areas Commun. **30**, 550–560 (2012)

15. A. Paulraj, R. Nabar, D. Gore, *Introduction to space-time wireless communications* (Cambridge University Press, 2003)
16. E.G. Larsson, O. Edfors, F. Tufvesson, T.L. Marzetta, Massive MIMO for next generation wireless systems. IEEE Commun. Mag. **52**, 186–195 (2014)
17. L. Lu, G.Y. Li, A.L. Swindlehurst, A. Ashikhmin, R. Zhang, An overview of massive MIMO: benefits and challenges. IEEE J. Sel. Topics Signal Process. **8**, 742–758 (2014)
18. D.J. Love, J. Choi, P. Bidigare, A closed-loop training approach for massive MIMO beamforming systems, in *2013 47th Annual Conference on Information Sciences and Systems (CISS)*, pp. 1–5, Mar. 2013
19. O. Elijah, C.Y. Leow, T.A. Rahman, S. Nunoo, S.Z. Iliya, A comprehensive survey of pilot contamination in massive MIMO–5G system. IEEE Commun. Surv. Tuts. **18**, 905–923 (2016) (2nd Quart.)
20. T.L. Marzetta, Noncooperative cellular wireless with unlimited numbers of base station antennas. IEEE Trans. Wirel. Commun. **9**, 3590–3600 (2010)
21. F. Rusek, D. Persson, B.K. Lau, E.G. Larsson, T.L. Marzetta, O. Edfors, F. Tufvesson, Scaling up MIMO: opportunities and challenges with very large arrays. IEEE Signal Process. Mag. **30**, 40–60 (2013)
22. J. Choi, D.J. Love, T. Kim, Trellis-extended codebooks and successive phase adjustment: A path from LTE-advanced to FDD massive MIMO systems. IEEE Trans. Wirel. Commun. **14**, 2007–2016 (2015)
23. T.S. Rappaport, R.W. Heath Jr., R.C. Daniels, J.N. Murdock, *Millimeter wave wireless communications* (Pearson Education, 2014)
24. S. Hur, T. Kim, D.J. Love, J.V. Krogmeier, T.A. Thomas, A. Ghosh, Millimeter wave beamforming for wireless backhaul and access in small cell networks. IEEE Trans. Commun. **61**, 4391–4403 (2013)
25. E. Torkildson, U. Madhow, M. Rodwell, Indoor millimeter wave MIMO: Feasibility and performance. IEEE Trans. Wirel. Commun. **10**, 4150–4160 (2011)
26. M.R. Akdeniz, Y. Liu, M.K. Samimi, S. Sun, S. Rangan, T.S. Rappaport, E. Erkip, Millimeter wave channel modeling and cellular capacity evaluation. IEEE J. Sel. Areas Commun. **32**, 1164–1179 (2014)
27. O.E. Ayach, S. Rajagopal, S. Abu-Surra, Z. Pi, R.W. Heath Jr., Spatially sparse precoding in millimeter wave MIMO systems. IEEE Trans. Wirel. Commun. **13**, 1499–1513 (2014)
28. X. Yu, J. Shen, J. Zhang, K.B. Letaief, Alternating minimization algorithms for hybrid precoding in millimeter wave MIMO systems. IEEE J. Sel. Topics Signal Process. **10**, 485–500 (2016)
29. C.N. Barati, S.A. Hosseini, M. Mezzavilla, T. Korakis, S.S. Panwar, S. Rangan, M. Zorzi, Initial access in millimeter wave cellular systems. IEEE Trans. Wirel. Commun. **15**, 7926–7940 (2016)
30. Z. Gao, C. Hu, L. Dai, Z. Wang, Channel estimation for millimeter-wave massive MIMO with hybrid precoding over frequency-selective fading channels. IEEE Commun. Lett. **20**, 1259–1262 (2016)
31. T.S. Rappaport, *Wireless communications: principles and practice*, vol. 2. Prentice Hall PTR New Jersey, 1996
32. R. Nasri, A. Jaziri, Tractable approach for hexagonal cellular network model and its comparison to poisson point process, in *IEEE Global Communications Conference (GLOBECOM)*, pp. 1–6, Dec. 2015
33. A.D. Wyner, Shannon-theoretic approach to a Gaussian cellular multiple-access channel. IEEE Trans. Inf. Theor. **40**, 1713–1727 (1994)
34. O. Somekh, S. Shamai, Shannon-theoretic approach to a gaussian cellular multiple-access channel with fading. IEEE Trans. Inf. Theor. **46**, 1401–1425 (2000)
35. A.J. Viterbi, A.M. Viterbi, E. Zehavi, Other-cell interference in cellular power-controlled CDMA. IEEE Trans. Commun. **42**, 1501–1504 (1994)
36. J. Xu, J. Zhang, J.G. Andrews, On the accuracy of the Wyner model in cellular networks. IEEE Trans. Wirel. Commun. **10**, 3098–3109 (2011)

37. P. Gupta, P.R. Kumar, The capacity of wireless networks. IEEE Trans. Inf. Theor. **46**, 388–404 (2000)
38. M. Grossglauser, D.N.C. Tse, Mobility increases the capacity of ad hoc wireless networks. IEEE/ACM Trans. Netw. **10**, 477–486 (2002)
39. X. Descombes, *Stochastic geometry for image analysis* (John Wiley & Sons, 2013)
40. M. Haenggi, *Stochastic geometry for wireless networks* (Cambridge University Press, Cambridge, U.K., 2012)
41. F. Baccelli, B. Błaszczyszyn, *Stochastic geometry and wireless networks: volume I theory* (vol. 3. Now Publishers Inc., 2009)
42. A.M. Hunter, J.G. Andrews, S. Weber, Transmission capacity of ad hoc networks with spatial diversity. IEEE Trans. Wirel. Commun. **7**, 5058–5071 (2008)
43. S.P. Weber, X. Yang, J.G. Andrews, G. de Veciana, Transmission capacity of wireless ad hoc networks with outage constraints. IEEE Trans Inf. Theory **51**, 4091–4102 (2005)
44. W.C. Cheung, T.Q.S. Quek, M. Kountouris, Throughput optimization, spectrum allocation, and access control in two-tier femtocell networks. IEEE J. Sel. Areas Commun. **30**, 561–574 (2012)
45. T.D. Novlan, R.K. Ganti, A. Ghosh, J.G. Andrews, Analytical evaluation of fractional frequency reuse for OFDMA cellular networks. IEEE Trans. Wirel. Commun. **10**, 4294–4305 (2011)
46. C. Li, J. Zhang, K.B. Letaief, Throughput and energy efficiency analysis of small cell networks with multi-antenna base stations. IEEE Trans. Wirel. Commun. **13**, 2505–2517 (2014)
47. C. Li, J. Zhang, J.G. Andrews, K.B. Letaief, Success probability and area spectral efficiency in multiuser MIMO HetNets. IEEE Trans. Commun. **64**, 1544–1556 (2016)
48. C. Li, J. Zhang, M. Haenggi, K.B. Letaief, User-centric intercell interference nulling for downlink small cell networks. IEEE Trans. Commun. **63**, 1419–1431 (2015)
49. X. Yu, C. Li, J. Zhang, K.B. Letaief, A tractable framework for performance analysis of dense multi-antenna networks, in *Proceedings of IEEE International Conference Communications (ICC)*, (Paris, France), pp. 1–6, May 2017
50. X. Yu, C. Li, J. Zhang, M. Haenggi, K.B. Letaief, A unified framework for the tractable analysis of multi-antenna wireless networks. IEEE Trans. Wirel. Commun. **17**, 7965–7980 (2018)
51. S. Weber, J.G. Andrews, *Transmission capacity of wireless networks* (vol. 5. Now Publishers Inc., 2012)
52. H.H. Yang, T.Q. Quek, *Massive MIMO meets small cell: backhaul and cooperation* (Springer, 2017)

Chapter 2
Fundamentals of Wireless Network Analysis

Abstract In this chapter, the fundamentals of wireless network analysis via stochastic geometry are introduced. The Poisson network model is first presented, and key performance metrics in wireless networks are defined. By modeling a wireless network as a Poisson point process, the distribution of the aggregate interference is characterized using the Laplace transform, which is a key analytical step leading to tractable results of the signal-to-interference-plus-noise ratio (SINR) distribution. Sample results are presented for coverage and rate analysis in single-antenna cellular and ad hoc networks.

2.1 Spatial Models of Wireless Networks

As the mobile data traffic keeps increasing at an exponential rate, mobile operators are deploying more and more access points, potentially with different types. Thus, the spatial distribution of the radio transceivers in modern wireless networks becomes more and more irregular, which makes the traditional network analysis based on the simplified hexagonal-shaped cellular model inapplicable. To analyze such complex wireless networks, a mathematical spatial model that is both tractable and accurate is required. In this section, we introduce basic concepts of stochastic geometry, of which the point process is the most important subfield. Then the Poisson network model is introduced, which will be the main network model in this monograph.

2.1.1 Stochastic Geometry

Stochastic geometry, originally called geometric probability, is the study of random spatial patterns, with which a collection of points in the Euclidean space can be mathematically and rigorously described. As the locations of nodes, or equivalently the network topology are random in nature due to the dynamic number and constant mobility of users, stochastic geometry stands out as an excellent candidate to model the spatial distribution of wireless networks.

X. Yu et al., *Stochastic Geometry Analysis of Multi-Antenna Wireless Networks*, https://doi.org/10.1007/978-981-13-5880-7_2

A stochastic geometry-based spatial model was initially developed to investigate wireless networks in the early 1960s [1], and later it was used in the 1970s and 1980s for examining packet radio networks [2–4]. Afterward, their use increased significantly for studying ad hoc networks, e.g., [2, 5–13]. The modeling of cellular networks by stochastic geometry first appeared in 1997 [14, 15], but the focus of those papers was not on coverage or throughput, but on the distribution of the number of users. In 2011, Andrews et al. [16] proposed a tractable approach to derive the coverage and throughput for cellular networks via stochastic geometry. These early papers have resulted in a flurry of followup works. Promising results in these works showed that wireless network analysis via stochastic geometry allows for the derivation of closed-form or semi-closed-form expressions for key network performance metrics without resorting to system-level simulations or (possibly intractable or inaccurate) deterministic models. Interested readers can refer to the textbooks and monographs such as [17–22], and survey papers such as [23].

The performance of wireless networks is fundamentally determined by co-channel interference. The topology of the network critically determines the distribution of the aggregated interference from other co-channel transmitters and therefore plays a fundamental role. This is quite different from the case in wired networks, where the underlying geometry is less important. In particular, the random objects in stochastic geometry are usually simple points, which represent the locations of network nodes such as receivers and transmitters, and the Euclidean space is either three-dimensional, or more often, the (two-dimensional) plane, which represents a geographical region. To model the spatial pattern of the points representing the transceivers in wireless networks, the point process, the heart of stochastic geometry theory, is of great value for detailed investigation. In the following, we shall first introduce basic point processes and apply them to build wireless network models.

2.1.2 Point Processes

The basic ingredients of practical geometry are points. A *point process* is a countable random collection of points that reside in some measure space, usually the Euclidean space $\mathbb{R}^d$ [21, Definition 2.1]. The simplest point process is the one where only one point appears. In particular, a random point x uniformly distributed in a compact set $W \subset \mathbb{R}^d$ is one point such that

$$\mathbb{P}(x \in A) = \frac{\nu_d(A)}{\nu_d(W)}, \tag{2.1}$$

where $\mathbb{P}(x \in A)$ is the probability that a random point x falls in an arbitrary set $A \subset W$, and $\nu_d(\cdot)$ is the Lebesgue measure[1] in the d-dimension space. We mainly

[1]The mathematical definition of Lebesgue measure is not provided here but can be well found in [21, 24], while there are some special cases for d that are easy to understand and will be further

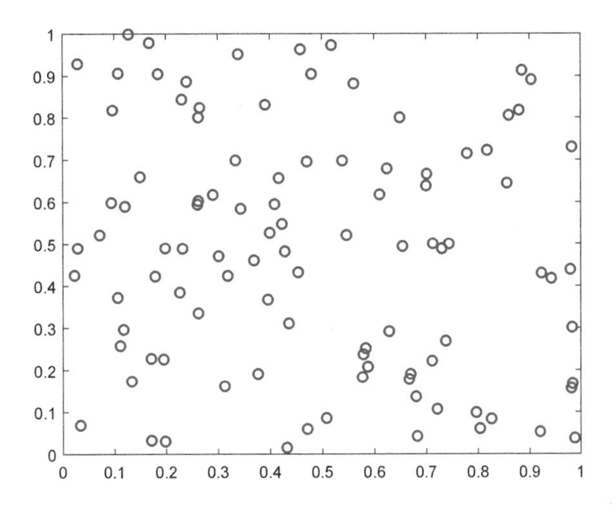

Fig. 2.1 An example of BPPs with $n = 100$ in $W = [0, 1]^2$

consider the two-dimension space. In stochastic geometry, many point processes are introduced on the basis of the uniformly distributed points. For example, a *binomial point process (BPP)* is a basic and well-known point process, and its definition is given as follows.

Definition 2.1 (*Binomial point processes*) A BPP of n points is defined by n independent points $x_1, x_2, \ldots, x_n$ that are uniformly distributed in a common compact set W, with the distribution given by

$$\begin{aligned}\mathbb{P}(x_1 \in A_1, x_2 \in A_2, \ldots, x_n \in A_n) &= \mathbb{P}(x_1 \in A_1)\mathbb{P}(x_2 \in A_2) \times \cdots \times \mathbb{P}(x_n \in A_n)\\ &= \frac{\prod_{i=1}^{n} \nu_d(A_i)}{[\nu_d(W)]^n}\end{aligned} \tag{2.2}$$

for subsets $A_1, A_2, \ldots, A_n \subset W$.

Figure 2.1 shows an example of BPPs with $n = 100$ in $W = [0, 1]^2$. We denote a BPP of n points as $\Phi_{W^{(n)}}$ and the number of points in any subset $A \subset W$ as $\Phi_{W^{(n)}}(A)$.

Note that $\Phi_{W^{(n)}}(A)$ has a binomial distribution with parameters n and $p = \nu_d(A)/\nu_d(W)$, and this is the reason why $\Phi_{W^{(n)}}$ is named after "binomial". Therefore, the expected value of $\Phi_{W^{(n)}}(A)$ is given by

$$\mathbb{E}\left[\Phi_{W^{(n)}}(A)\right] = \lambda \nu_d(A), \quad \text{for } A \subset W, \tag{2.3}$$

where

$$\lambda = \frac{n}{\nu_d(W)} \tag{2.4}$$

used in this monograph. When $d = 1$, $\nu_1(\cdot) = l$ is the length measure; when $d = 2$, $\nu_2(\cdot) = A$ is the area measure; when $d = 3$, $\nu_3(\cdot) = V$ is the volume measure.

is defined as the *intensity* of the BPP, representing the mean number of points per unit volume. An observation for BPP is that $\Phi_{W^{(n)}}(A_i)$ are not independent even though A_i's are disjoint subsets of W, which is mainly due to the fixed number of points in W. A simple example to illustrate the dependency is that $\Phi_{W^{(n)}}(W \backslash A) = n - m$ if $\Phi_{W^{(n)}}(A) = m$.

It has been shown in [21, Theorem 2.23] that a simple point process is completely characterized by its *void probability*. In other words, two simple point processes have the same distribution if their void probabilities are identical. The void probability is defined by the probability that there is no point in an arbitrary compact subset $K \subset W$. For BPPs, the void probability is given by

$$\mathbb{P}\left(\Phi_{W^{(n)}}(K) = 0\right) = \frac{[v_d(W) - v_d(K)]^n}{[v_d(W)]^n}. \tag{2.5}$$

Note that the void probability shall be frequently used in wireless network performance analysis, especially when characterizing the distribution of the distance between the receiver and its nearest transmitter.

2.1.3 Poisson Network Models

Recall that, for a BPP $\Phi_{W^{(n)}}$, the number of points falling in $A \subset W$, denoted as $\Phi_{W^{(n)}}(A)$, is a random variable distributed according to the binomial distribution with parameters n and $p = v_d(A)/v_d(W)$. The well-known Poisson limit theorem yields that $\Phi_{W^{(n)}}(A)$ is asymptotically Poisson distributed with the parameter $\lambda v_d(A)$, when $n \to \infty$ and $p \to 0$. The condition is satisfied by enlarging W to fill out the whole $\mathbb{R}^d$ while n tends to infinity, which is equivalent to relaxing the BPP to an unbounded case, and this limiting point process is called a Poisson point process (PPP). A formal definition is given below.

Definition 2.2 (*Homogeneous Poisson point processes*) A homogeneous PPP Φ is established if the following two conditions are satisfied:

- *Poisson distributed point counts*: The number of points in a bounded set A is Poisson distributed, i.e.,

$$\mathbb{P}\left(\Phi(A) = m\right) = \frac{\bar{\mu}^m}{m!} e^{-\bar{\mu}}, \quad \text{for } m = 0, 1, 2, \ldots, \tag{2.6}$$

  where $\bar{\mu} = \lambda v_d(A)$ is the mean of the Poisson distribution. The parameter λ is called the intensity or *density* of the homogeneous PPP.
- *Independent scattering*: The number of points $\{\Phi(A_i)\}_{i=1}^p$ in p disjoint subsets $\{A_i\}_{i=1}^p$ are p independent random variables.

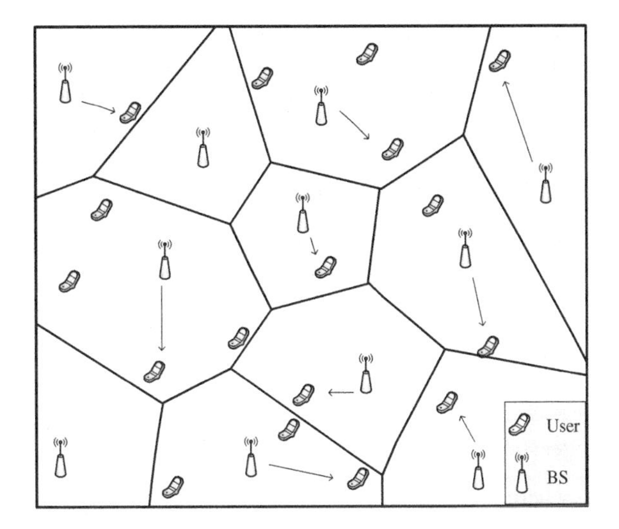

Fig. 2.2 An example of the cellular network model, where the BSs and users are distributed according to two independent homogeneous PPPs. ©2014 IEEE. Reprinted, with permission, from [25]

Once the density λ is determined, the distribution of a homogeneous PPP is determined. According to (2.6), the void probability for a homogeneous PPP is given by

$$\mathbb{P}(\Phi(K) = 0) = \exp\{-\lambda v_d(K)\}. \tag{2.7}$$

As indicated before, the homogeneous PPP is closely related to BPP. In particular, by restricting a homogeneous PPP to a compact set W conditioned on there are n points in W, one can obtain a BPP in W with n points. This assertion can be proved by considering the voiding probabilities of these two point processes. If $K \subset W$, then the voiding probability of the conditioned homogeneous PPP is written by

$$\begin{aligned}
\mathbb{P}(\Phi(K) = 0|\Phi(W) = n) &= \frac{\mathbb{P}(\Phi(K) = 0, \Phi(W) = n)}{\mathbb{P}(\Phi(W) = n)} \\
&= \frac{\mathbb{P}(\Phi(K) = 0)\mathbb{P}(\Phi(W \backslash K) = n)}{\mathbb{P}(\Phi(W) = n)} \\
&= \frac{[v_d(W) - v_d(K)]^n}{[v_d(W)]^n},
\end{aligned} \tag{2.8}$$

which is exactly the same as the void probability $\mathbb{P}\left(\Phi_{W^{(n)}}(K) = 0\right)$ of the BPP $\Phi_{W^{(n)}}$ in (2.5), and hence indicates that these two point processes are closely related.

The homogeneous PPP has been widely used to model the spatial distribution of nodes in wireless networks [16, 18], where the transceivers are uniformly distributed in an infinite area. Figure 2.2 gives one realization of a PPP modeled cellular network, where the BSs and users are distributed according to two independent homogeneous PPPs. Recent investigations have demonstrated that, with the homogeneous PPP model, the performance analysis of wireless networks becomes tractable. Subsequently, network design insights can be analytically revealed, which will be illustrated in Sect. 2.4 for single-antenna wireless networks.

In a homogeneous PPP, its density λ is an invariant constant, which means that the average number of points per unit area does not change over the space. Nevertheless, the density of the point process may vary with the positions in some applications. For example, the density of the transceivers in wireless networks may vary according to the density of population, which could be different in urban or rural areas. This motivates the consideration of another kind of PPPs, called inhomogeneous PPPs.

Definition 2.3 (*Inhomogeneous Poisson point processes*) An inhomogeneous PPP satisfies the following two properties:

- *Poisson distributed point counts*: The number of points in a bounded set A is Poisson distributed, i.e.,

$$\mathbb{P}\left(\Phi(A)=m\right)=\frac{\Lambda(A)^m}{m!}e^{-\Lambda(A)},\quad \text{for } m=0,1,2,\ldots, \tag{2.9}$$

  where

$$\Lambda(A)=\int_A \lambda(x)\mathrm{d}x. \tag{2.10}$$

  The spatially variant density is denoted as $\lambda(x)$ and is a function of the location x in the space, while $\Lambda(A)$ is the intensity measure of the set A representing the expected number of points in A.
- *Independent scattering*: The number of points $\{\Phi(A_i)\}_{i=1}^p$ in p disjoint subsets $\{A_i\}_{i=1}^p$ are p independent random variables.

According to (2.9), the void probability for an inhomogeneous PPP is given by

$$\mathbb{P}(\Phi(K)=0)=\exp\left\{-\int_A \lambda(x)\mathrm{d}x\right\}. \tag{2.11}$$

Another commonly encountered point process for wireless networks is the Poisson hole process.[2] The definition of the 2D Poisson hole processes is given below.

Definition 2.4 (*Poisson hole processes*) Let $\Phi_1, \Phi_2 \subset \mathbb{R}^2$ be two independent homogeneous PPPs with densities λ_1 and λ_2, respectively. Define a set

$$\Xi_r \triangleq \bigcup\{x\in\Phi_1 : b(x,r)\}, \tag{2.12}$$

which is the union of all disks of radius r centered at a point of Φ_1. The Poisson hole process Φ is then defined by

$$\Phi=\{x\in\Phi_2 : x\notin\Xi_r\}=\Phi_2\backslash\Xi_r, \tag{2.13}$$

which is composed of the points in Φ_2 but out of the disks of radius r from Φ_1.

---

[2] Here we pick the two-dimensional Poisson hole process as the example for the ease of presentation.

The Poisson hole process Φ is basically a Cox process [21, Chap. 3.3] with measure $\lambda_2 \nu_d(\cdot \backslash \Xi_r)$. It is widely used for modeling cognitive radio networks, where Φ_1 represents the primary users while Φ_2 denotes the secondary users [26]. In this way, the Poisson hole process Φ can be interpreted as the secondary users that are allowed to access the channel. The probability that a point is preserved is the probability that there is no point falling in Φ_1 within radius r. Therefore, the density of the process Φ is

$$\lambda = \lambda_2 \exp(-\lambda_1 \pi r^2). \tag{2.14}$$

While there are many other kinds of point processes available in the theory of stochastic geometry, in the remaining parts of this monograph, we will mainly focus on PPPs due to their tractability and accuracy for modeling modern wireless networks. There is one important theorem called Slivnyak's theorem for PPPs, and it is given as follows.

Theorem 2.1 (Slivnyak's Theorem) *For a PPP Φ, the following two probabilities are the same:*

$$\mathbb{P}\left(\Phi \backslash \{x\} \in Y \mid x \in \Phi\right) \equiv \mathbb{P}\left(\Phi \in Y\right), \tag{2.15}$$

where Y is an event that can be viewed as a property of the point process.

The proof of the Slivnyak's theorem can be found in [21, Chap. 8]. Note that the Slivnyak's theorem suggests that conditioning on a point at x does not change the distribution of the rest of the process. In addition, the Slivnyak's theorem is invertible, i.e., if the condition in (2.15) is satisfied for a point process, it must be a PPP. The Slivnyak's theorem has an important implication in analyzing wireless networks based on PPPs. Specifically, it implies that conditioning on $x \in \Phi$ in a PPP is the same as adding a point at x, and this can be extended to multiple points.

A *typical point* is usually selected as the object in the performance analysis of wireless networks. It results from a selection procedure in which every point has the same chance of being selected. According to Slivnyak's theorem, to analyze the performance of the typical point, we can artificially deploy it at the origin of the space, which is an analytical trick in many studies. In this way, we consider the point at the origin, which, under an expectation over the point process, becomes the *typical receiver*.

2.1.4 Examples of Poisson Network Models

In this subsection, we present two examples of the Poisson network model to illustrate the main ingredients, namely the cellular network model and the ad hoc network model. Here we consider single-antenna networks, while multi-antenna network models will be introduced in Chap. 3.

Cellular Networks

A cellular network mainly consists of BSs and users, as shown in Fig. 2.2. For a downlink cellular network, the BSs transmit radio signals to their served users. To analyze the network performance, the BSs are commonly modeled as a homogeneous PPP in $\mathbb{R}^2$, denoted by Φ. The density of the BSs is denoted by λ. Consider the typical user located at the origin, and usually, it is served by the nearest BS located at x_0 with distance r_0. The signal power received by the typical user is given by

$$S = P_{\mathrm{t}} g_{x_0} L\left(r_0\right), \tag{2.16}$$

where P_{t} is the transmit power of the serving BS. And g_{x_0} is the channel power gain from the BS to the typical user, which is random and depends on the channel fading distribution. In addition, $L\left(r_0\right)$ is the path loss function, which is basically a decreasing function with respect to the distance r_0. There are several commonly used path loss models. For example,

$$\begin{cases} L\left(d\right) = d^{-\alpha}, & (a) \\ L\left(d\right) = \left(1+d\right)^{-\alpha}, & (b) \\ L\left(d\right) = \left(\max\left(d_0, d\right)\right)^{-\alpha}, & (c) \end{cases} \tag{2.17}$$

where $\alpha > 2$ is called the path loss exponent. (a) is the standard power law path loss model, which is the most commonly adopted model, but it is not applicable when $d < 1$. Therefore, (b) and (c) are proposed to fix this issue where $d_0 \geq 1$.

Due to the broadcast nature of wireless communications, the typical user not only receives the signal from its associated BS but also receives interference from other BSs transmitting on the same radio channel. The interference power received by the typical user is given by

$$I = \sum_{x \in \Phi \setminus \{x_0\}} P_x g_x L\left(\|x\|\right), \tag{2.18}$$

where P_x is the transmit power of the BS at x, g_x is the channel gain between the BS at x and the typical user, and $\|x\|$ represents the distance from the BS at x to the typical user at the origin. Therefore, the signal-to-interference-plus-noise ratio (SINR) of the typical user is given by

$$\mathrm{SINR} = \frac{S}{I+N} = \frac{P_{\mathrm{t}} g_{x_0} L\left(r_0\right)}{\sum_{x \in \Phi \setminus \{x_0\}} P_x g_x L\left(\|x\|\right) + N}, \tag{2.19}$$

where N denotes the white Gaussian noise power.

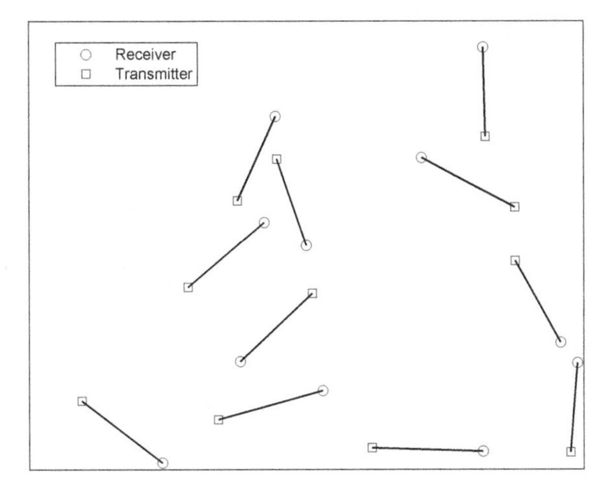

Fig. 2.3 An example of the ad hoc network model

Ad Hoc Networks

In an ad hoc network, each transmitter has an associated receiver. Figure 2.3 shows an illustration. Ad hoc networks are typically modeled by Poisson bipolar models. Specifically, in a Poisson bipolar network, all the transmitters are modeled as a PPP, with a set of receivers at fixed distance r_0 and uniformly and randomly chosen orientations from their transmitters. Hence, by the displacement theorem [21, Theorem 2.33], a Poisson bipolar network is composed of two dependent PPPs for transmitters and receivers, respectively.

Assume the typical receiver located at the origin in $\mathbb{R}^2$, and its associated transmitter is located at x_0 with a fixed distance r_0. The point process that describes the interfering transmitters consists of the whole PPP Φ except x_0. According to the Slivnyak's theorem, the distribution of $\Phi \backslash \{x_0\}$ is the same as Φ. Therefore, the SINR of the typical receiver in ad hoc networks is expressed as

$$\mathrm{SINR} = \frac{S}{I+N} = \frac{P_{\mathrm{t}} g_{x_0} L\left(r_0\right)}{\sum_{x \in \Phi} P_x g_x L\left(\|x\|\right) + N}, \tag{2.20}$$

which is almost the same as that of the typical user in a cellular network. However, there is one major difference compared with the cellular network model. In cellular networks, the interfering BSs are farther away than r_0, if the typical user is served by the nearest BS. Even in heterogeneous cellular networks (which will be discussed in Chap. 5), the distances from the interfering BSs to the typical user depend on r_0. On the contrary, in ad hoc networks, the locations of the interfering transmitters do not depend on r_0, which means that the interfering transmitters could be arbitrarily close to the typical receiver. Such a difference in the spatial distribution of interference will cause a big difference in the analysis, as well as behavior, of cellular networks and ad hoc networks, as will be illustrated later.

2.2 Network Performance Metrics

In this section, we introduce major network performance metrics that will be considered in this monograph. A receiver can recover the information from its transmitter only if it can distinguish the desired signal from the interference and noise. To evaluate the signal quality, the SINR is the main performance indicator in wireless communications. Most network performance metrics are determined by the SINR at the receiver. In the following, we will present coverage probability, spectral efficiency, and energy efficiency, respectively.

Coverage Probability and Outage Probability

Coverage probability is a fundamental performance metric for a wireless network. It is also called the *success probability*, which measures the reliability of a transmission link. The definition of the coverage probability is given by

$$p_{\mathrm{c}} = \mathbb{P}\left(\mathrm{SINR} \geq \tau\right), \tag{2.21}$$

where τ is the SINR threshold. The coverage probability can be viewed from different aspects. First, as we mentioned before, it measures the reliability of a transmission link. Second, it also represents the average fraction of receivers which achieve the SINR threshold at any time. Third, it is the average fraction of the network area which is under "coverage" at any time. Lastly, from (2.21) it is shown that the coverage probability is exactly the complementary cumulative distribution function (ccdf) of the SINR.

Outage probability, on the contrary, measures the probability that the transmission of a link fails. It is given by

$$p_{\mathrm{o}} = 1 - p_{\mathrm{c}} = \mathbb{P}\left(\mathrm{SINR} < \tau\right). \tag{2.22}$$

Spectral Efficiency

Spectral efficiency is another important performance metric of wireless networks. It measures the "transmission speed" of a network. For a fixed-rate transmission, the *spectral efficiency* of a communication link is defined as

$$R_{\mathrm{t}} = p_{\mathrm{c}} R_0, \tag{2.23}$$

where p_{c} is given in (2.21), and R_0 is the achievable data rate per link normalized by the bandwidth. If each transmitter serves one receiver, R_0 is written as

$$R_0 = \log_2\left(1 + \tau\right). \tag{2.24}$$

If each transmitter serves multiple receivers simultaneously, e.g., by MIMO multiplexing techniques, then R_0 is given as

$$R_0 = U \log_2 (1 + \tau), \tag{2.25}$$

where U is the number of receivers served by the transmitter. The units of R_0 and R_t are bit/s/Hz.

From the network point of view, it is more important to know the spectral efficiency of a unit area, which is called the *area spectral efficiency* (ASE). The definition of the ASE is given by

$$R_a = \lambda_t R_t = \lambda_t p_c R_0, \tag{2.26}$$

where λ_t is the density of the active transmitters in the area. The active transmitters are the ones that send signals to their receivers simultaneously. If the density of the transmitters is λ and each of them has the probability p_a to be active, then $\lambda_t = \lambda p_a$. In addition, the unit of ASE is bit/s/Hz/m^2.

Note that the spectral efficiency and ASE discussed above is for fixed-rate transmissions. In some literature, the achievable rate per user with rate adaptation, defined as $\mathbb{E}\left[\log_2 (1 + \text{SINR})\right]$, is also used as a performance metric. This metric is also determined by the SINR distribution, i.e., the coverage probability. Its analysis will not be considered in this monograph, and we mainly focus on the spectral efficiency and ASE.

Energy Efficiency

Energy efficiency is a metric that has become increasingly important and received significant attention in recent years. It is defined as the ratio of the ASE to the power consumption per unit area. Mathematically, the *energy efficiency* is defined as

$$\xi_{\text{EE}} = \frac{R_a}{P_a}, \tag{2.27}$$

where P_a is the power consumption per unit area. The unit of the energy efficiency is bits/J/Hz. For a wireless network, the power consumption includes the consumption of both transmitters and receivers. In cellular networks, however, the power consumption mainly counts the consumption from the BSs, as they consume the largest portion of the overall energy [27].

The power consumption of a BS can be modeled as [28, 29]

$$P_{\text{BS}} = \frac{1}{\xi} P_t + M P_c + P_0, \tag{2.28}$$

where ξ is the power amplifier efficiency, P_t is the transmit power, M is the number of transmit antennas, and P_c accounts for the circuit power consumption which is proportional to M. In particular, P_c includes the power consumption of the corresponding RF chain. P_0 is the non-transmission power, which is the power consumption that is independent of both M and P_t. The units of P_t, P_c, P_0, and P_{BS} are watt. For the conventional macrocell, the non-transmission power consumes a large portion of the BS power. While for small cells, such as the femtocell, the proportion of power consumption from non-transmission part is much reduced.

The average power consumption per unit area includes the transmit power and circuit power consumption from active BSs, as well as the non-transmission power consumption from both active and inactive BSs, and it is given by

$$P_a = \lambda p_a \left(\frac{1}{\xi} P_t + M P_c \right) + \lambda P_0, \tag{2.29}$$

where λ is the density of the BS, and p_a is the BS active probability, defined as the ratio of the number of active BSs to the total number of BSs. Therefore, the energy efficiency of a cellular network is written as

$$\xi_{EE} = \frac{R_a}{P_a} = \frac{p_a p_c R_0}{p_a \left(\frac{1}{\xi} P_t + M P_c \right) + P_0}. \tag{2.30}$$

2.3 Interference Distribution in Poisson Wireless Networks

As shown in Sect. 2.2, the main network performance metrics are derived from the coverage probability, or, more precisely, the SINR distribution, for which, the distribution of the cumulated interference needs to be first characterized. In this section, we analyze the interference distribution, and the analysis of the SINR distribution will be presented in the next section.

In a general wireless network, the interference power received by a receiver at the origin is given by

$$I = \sum_{x \in \Phi'} P_x g_x L\left(\|x\|\right), \tag{2.31}$$

where Φ' is the set of the location of the interfering transmitters, and all the other parameters have been introduced in Sect. 2.1.4.

2.3.1 *Sums and Products over Point Processes*

To derive the distribution of the interference power in (2.31), the sums and products over point processes are typically encountered. In this subsection, we first introduce the distributions of the sums and products over point processes.

Let $\Phi = \{x_i\} \subset \mathbb{R}^d$ be a point process, and $f : \mathbb{R}^d \mapsto \mathbb{R}$ be a measurable function. A sum of $f(x)$ over the point process Φ is defined as

$$S[f] \triangleq \sum_{x \in \Phi} f(x). \tag{2.32}$$

Assuming Φ is a point process with density $\lambda(x)$, the expected value of the sum of $f(x)$ is given by Campbell's theorem.

Theorem 2.2 (Campbell's theorem for sums) *Let $f(x) : \mathbb{R}^d \rightarrow [0, \infty]$ be a measurable function and Φ is a point process with density $\lambda(x)$. Then*

$$\mathbb{E}\left[\sum_{x \in \Phi} f(x)\right] = \int_{\mathbb{R}^d} f(x)\lambda(x)\mathrm{d}x. \tag{2.33}$$

Campbell's theorem provides a way to calculate the sum over a point process. The *probability generating functional (PGFL)* presented in the following gives the method to calculate the product over a point process, which is widely used in the analysis of Poisson network models.

The PGFL of a point process Φ is defined by the expected value of the product of measurable functions $v : \mathbb{R}^d \mapsto [0, 1]$ over a point process Φ, written by

$$G[v] \triangleq \mathbb{E}\left(\prod_{x \in \Phi} v(x)\right) = \mathbb{E}\left[\exp\left(\sum_{x \in \Phi} \log v(x)\right)\right]. \tag{2.34}$$

This equation reveals the relation between the PGFL and a sum over the point process, i.e., from the definition in (2.32), we have

$$G[v] \equiv \mathbb{E}\left[e^{S[\log v]}\right], \tag{2.35}$$

which has a form of a *moment-generating function*. Using the definition of the moment-generating function of a random variable X, given by

$$M_X(t) \triangleq \mathbb{E}\left[e^{tX}\right], \quad t \in \mathbb{R}, \tag{2.36}$$

the PGFL is given by $G[v] = M_{S[\log v]}(1)$. In another way, using the definition of the Laplace transform of a random variable X, given by

$$\mathscr{L}_X(s) \triangleq \mathbb{E}\left[e^{-sX}\right], \quad s \in \mathbb{C}, \tag{2.37}$$

the PGFL is written as $G[v] = \mathscr{L}_{S[\log v]}(-1)$.

Next, we shall present the PGFL of a homogeneous PPP, which is commonly adopted to model the spatial distribution of wireless networks.

Theorem 2.3 (PGFL of a homogeneous PPP) *The PGFL of a homogeneous PPP with density λ is given by*

$$G[v] \triangleq \mathbb{E}\left(\prod_{x\in\Phi} v(x)\right) = \exp\left(\lambda \int_{\mathbb{R}^d} (v(x) - 1)\,\mathrm{d}x\right). \tag{2.38}$$

Proof Let $b(o, R)$ be the ball of radius R centered at the origin, and $n = \Phi\,(b(o, R))$ be the number of points in $b(o, R)$. In this way, we have

$$\mathbb{E}\left(\prod_{x\in\Phi} v(x)\right) = \lim_{R\to\infty} \mathbb{E}\left(\prod_{x\in\Phi\cap b(o,R)} v(x)\right). \tag{2.39}$$

First, we calculate the expectation conditioned on the fact that there are n points in the ball, which is given by

$$\mathbb{E}\left(\prod_{x\in\Phi\cap b(o,R)} v(x)\ \middle|\ n = \Phi\,(b(o, R))\right) = \left(\int_{b(o,R)} \frac{v(x)}{|b(o, R)|}\mathrm{d}x\right)^n. \tag{2.40}$$

Then, by averaging over n, the expectation $\mathbb{E}\left(\prod_{x\in\Phi\cap b(o,R)} v(x)\right)$ can be derived as

$$\begin{aligned}
\mathbb{E}\left(\prod_{x\in\Phi\cap b(o,R)} v(x)\right) &= \sum_{n=0}^{\infty}\left(\int_{b(o,R)} \frac{v(x)}{|b(o, R)|}\mathrm{d}x\right)^n \mathbb{P}\,(\Phi\,(b(o, R)) = n) \\
&= \sum_{n=0}^{\infty}\left(\int_{b(o,R)} \frac{v(x)}{|b(o, R)|}\mathrm{d}x\right)^n \frac{(\lambda\,|b(o, R)|)^n}{n!} \exp\left(-\lambda\,|b(o, R)|\right) \\
&= \exp\left(-\lambda\,|b(o, R)| + \lambda \int_{b(o,R)} v(x)\mathrm{d}x\right) \\
&= \exp\left(-\lambda \int_{b(o,R)} [1 - v(x)]\,\mathrm{d}x\right).
\end{aligned} \tag{2.41}$$

By taking the limit in (2.39), the theorem is proved.

In this way, adopting the relation between the PGFL and moment-generating function, the moment-generating function of PPP is given in the following corollary.

Corollary 2.1 (Campbell's theorem of a homogeneous PPP) *Let Φ be a homogeneous PPP of intensity λ on $\mathbb{R}^d$. The moment-generating function of the sum $S = \sum_{x\in\Phi} f(x)$ is given by*

$$\mathbb{E}\left[e^{tS}\right] = \exp\left(\lambda \int_{\mathbb{R}^d} \left(e^{tf(x)} - 1\right)\mathrm{d}x\right) \tag{2.42}$$

for any complex t with which the integral converges.

Theorem 2.3 and Corollary 2.1 will be commonly used to derive various performance metrics for wireless networks throughout this monograph.

2.3.2 *Laplace Transform of Interference Distribution*

In this subsection, we assume the interfering transmitters form a homogeneous PPP, which means that all the interfering transmitters use the same transmit power and the channel gains from the transmitters to the typical receiver are identical. Without loss of generality, we normalize the transmit power P_x as 1. Thus, (2.31) can be written as

$$I = \sum_{x \in \Phi'} g_x L\left(\|x\|\right). \tag{2.43}$$

The Laplace transform of the interference completely characterizes the distribution of the interference, and it is a key function to derive the distribution of the SINR. We focus on the Laplace transform of the interference in (2.43) in this subsection.

The Laplace transform of interference given in (2.43) is given by

$$\mathscr{L}_I\left(s\right) = \mathbb{E}\left[e^{-sI}\right] = \mathbb{E}\left[e^{-s\sum_{x \in \Phi'} g_x L(\|x\|)}\right]. \tag{2.44}$$

Since g_x are independently distributed, $\mathscr{L}_I\left(s\right)$ in the above equation can be further written as

$$\mathscr{L}_I\left(s\right) = \mathbb{E}\left[\prod_{x \in \Phi'} e^{-sg_x L(\|x\|)}\right] = \mathbb{E}_{\Phi'}\left[\prod_{x \in \Phi'} \mathbb{E}_{g_x}\left[e^{-sg_x L(\|x\|)}\right]\right]. \tag{2.45}$$

Based on the PGFL in (2.38), (2.45) can be expressed as

$$\mathscr{L}_I\left(s\right) = \exp\left\{-\int_{\mathbb{R}^d}\left(1 - \mathbb{E}_{g_x}\left[e^{-sg_x L(\|x\|)}\right]\right)\Lambda\left(\mathrm{d}x\right)\right\}, \tag{2.46}$$

where $\Lambda\left(\mathrm{d}x\right)$ is the intensity measure function at $\mathrm{d}x$. For a homogeneous PPP with density λ, $\Lambda\left(\mathrm{d}x\right) = \lambda v_d\left(\mathrm{d}x\right)$.

2.3.3 *Example: Rayleigh Fading*

The Rayleigh fading channel is one of the most commonly adopted channel models in wireless communications. The Rayleigh fading is a statistical model for the effect of the multipath environment on the radio signals. It is ideally suited to situations where there are large numbers of signal paths and reflections, such as cellular networks where there are a large number of reflections from buildings. For single-antenna transmissions with Rayleigh fading channels, the channel gain g_x is exponentially distributed, i.e., $g_x \sim \text{Exp}(1)$. We consider a 2D PPP and use the standard power law path loss model $L(d) = d^{-\alpha}$. Then the Laplace transform of interference in (2.46) can be written as

$$\mathcal{L}_I(s) = \exp\left\{-2\pi\lambda \int_{l(r_0)}^{\infty} \left(1 - \mathbb{E}_{g_x}\left[e^{-sg_x r^{-\alpha}}\right]\right) r\mathrm{d}r\right\} \tag{2.47}$$

$$= \exp\left\{-2\pi\lambda \int_{l(r_0)}^{\infty} \left(1 - \frac{1}{1+sr^{-\alpha}}\right) r\mathrm{d}r\right\} \tag{2.48}$$

$$= \exp\left\{-\pi\lambda \int_{l(r_0)^2}^{\infty} \left(\frac{sv^{-\frac{\alpha}{2}}}{1+sv^{-\frac{\alpha}{2}}}\right) \mathrm{d}v\right\}, \tag{2.49}$$

where (2.47) follows the path loss model $L(\|x\|) = \|x\|^{-\alpha}$ and $r = \|x\|$, (2.48) is based on the exponential distribution of g_x, and (2.49) uses the substitution $v = r^2$. The lower limit of the integral $l(r_0)$ is the minimal distance between the interfering transmitters and the typical receiver, which is a function of r_0, denoted by $l(r_0)$. For example, in ad hoc networks, the interfering transmitters could be as close as possible, i.e., $l(r_0) = 0$. In cellular networks, when the typical user is served by the nearest BS, then $l(r_0) = r_0$, where r_0 is the distance between the typical user and its associated BS. Note that when $l(r_0) = 0$, the Laplace transform of the interference in (2.49) can be further expressed as

$$\mathcal{L}_I(s) = \exp\left\{-\pi\lambda s^{\delta} \frac{\pi\delta}{\sin(\pi\delta)}\right\}, \tag{2.50}$$

where $\delta \triangleq \frac{2}{\alpha}$. Note that the above equality holds when $\alpha > 2$.

2.4 Performance Analysis of Single-Antenna Networks

In this section, we continue using Rayleigh fading channels to illustrate the performance analysis of single-antenna networks.

Assume the receiver locates at the origin of a 2D plane. Its associated transmitter is with distance r_0. There are interfering transmitters on the plane, and they follow a homogeneous PPP Φ' with density λ. Assume all the transmitters use the same transmit power P. Then the SINR of the receiver is given by

$$\text{SINR} = \frac{S}{I+N} = \frac{Pg_{x_0}r_0^{-\alpha}}{N + \sum_{x\in\Phi'} Pg_x \|x\|^{-\alpha}}. \tag{2.51}$$

Denote $\sigma_N^2 = N/P$ as the normalized noise power, then the SINR is written as

$$\mathrm{SINR} = \frac{g_{x_0} r_0^{-\alpha}}{\sigma_N^2 + \sum_{x \in \Phi'} g_x \|x\|^{-\alpha}}. \tag{2.52}$$

The coverage probability is then given by

$$p_c = \mathbb{P}(\mathrm{SINR} \geq \tau) = \mathbb{P}\left(\frac{g_{x_0} r_0^{-\alpha}}{\sigma_N^2 + \sum_{x \in \Phi'} g_x \|x\|^{-\alpha}} \geq \tau\right). \tag{2.53}$$

To derive the coverage probability, one way is to use the distribution of g_{x_0}, which is $g_{x_0} \sim \exp(1)$ when the channel is Rayleigh fading. Therefore, the coverage probability can be written as

$$p_c = \mathbb{P}\left(g_0 \geq \tau r_0^{\alpha}\left(\sigma_N^2 + \sum_{x \in \Phi'} g_x \|x\|^{-\alpha}\right)\right) = \mathbb{E}\left[e^{-\tau r_0^{\alpha}\left(\sigma_N^2 + \sum_{x \in \Phi'} g_x \|x\|^{-\alpha}\right)}\right]. \tag{2.54}$$

By denoting $s = \tau r_0^{\alpha}$, then p_c can be expressed as

$$p_c = \mathbb{E}_{r_0}\left[e^{-s\sigma_N^2} \mathscr{L}_I(s)\right], \tag{2.55}$$

where $\mathscr{L}_I(s) = \mathbb{E}_I\left[e^{-sI} | r_0\right]$ is the Laplace transform of interference conditioned on the distance r_0, and it can be derived following Sect. 2.3. Next, we will present the SINR distribution and network performance of cellular networks and ad hoc networks, respectively.

2.4.1 *Cellular Networks*

Consider a downlink cellular network with Rayleigh fading channels. All the BSs and users are equipped with a single antenna. The system model is described in Sect. 2.1.4. Then the SINR distribution of the typical user is given in (2.55). To obtain the SINR distribution, we look into $\mathscr{L}_I(s)$ first. By substituting $s = \tau r_0^{\alpha}$ into (2.49), we have

$$\mathscr{L}_I\left(\tau r_0^2\right) = \exp\left\{-\pi\lambda \int_{l(r_0)^2}^{\infty} \left(\frac{\tau r_0^{\alpha} v^{-\frac{\alpha}{2}}}{1 + \tau r_0^{\alpha} v^{-\frac{\alpha}{2}}}\right) \mathrm{d}v\right\}. \tag{2.56}$$

Since the user is served by the nearest BS, all the interfering BSs are farther than r_0. Thus, in this case, $l(r_0) = r_0$. Then, let $\tau r_0^{\alpha} v^{-\frac{\alpha}{2}} \to u^{-\frac{\alpha}{2}}$, the Laplace transform of interference is given by

$$\mathscr{L}_I\left(\tau r_0^2\right) = \exp\left\{-\pi\lambda r_0^2 \tau^{\delta} \int_{\tau^{-\delta}}^{\infty} \frac{u^{-\frac{\alpha}{2}}}{1 + u^{-\frac{\alpha}{2}}} \mathrm{d}u\right\}. \tag{2.57}$$

Denote $c = \tau^{\delta} \int_{\tau^{-\delta}}^{\infty} \frac{u^{-\frac{\alpha}{2}}}{1+u^{-\frac{\alpha}{2}}} \mathrm{d}u$ and substitute (2.57) into (2.55), and then the coverage probability is given by

$$p_{\mathrm{c}} = \mathbb{E}_{r_0} \left[\exp \left(-\tau r_0^{\alpha} \sigma_N^2 - \pi \lambda c_0 r_0^2 \right) \right]. \tag{2.58}$$

Since the typical user is served by the nearest BS at distance r_0, it means that there is no other BS in the circle at the origin with radius r_0. Using the void probability of a PPP, the complementary cdf of r_0 is written as

$$\bar{F}(r_0) = \mathbb{P}\left(\text{No BS is in the area } \pi r_0^2\right) = e^{-\lambda \pi r_0^2}. \tag{2.59}$$

Then, the probability density function of r_0 is given by

$$f_{r_0}(r) = 2\pi \lambda r e^{-\pi \lambda r^2}. \tag{2.60}$$

Therefore, the coverage probability of the typical user is finally given by

$$p_{\mathrm{c}} = \int_0^{\infty} \pi \lambda \exp \left(-\tau v^{\frac{\alpha}{2}} \sigma_N^2 - \pi \lambda (1 + c_0) v \right) \mathrm{d}v. \tag{2.61}$$

Note that the performance analysis of the uplink transmission in cellular networks follows the same procedure. In the uplink model, the SINR at the BS is evaluated, in which the interference comes from the mobile users in other cells. The derivation also involves the Laplace transform of the interference distribution and application of the PGFL. Interested readers can refer to [30, 31] for detailed information.

Interference-Limited Networks

In wireless networks, it is reasonable to assume an interference-limited scenario, since the noise part σ_N^2 is very small compared with the interference part. By ignoring the noise power, the coverage probability can be derived from (2.61) to

$$p_{\mathrm{c}} = \frac{1}{1+c}. \tag{2.62}$$

Recall $c = \tau^{\delta} \int_{\tau^{-\delta}}^{\infty} \frac{u^{-\frac{\alpha}{2}}}{1+u^{-\frac{\alpha}{2}}} \mathrm{d}u$, and it can be represented by the hypergeometric function [32, Sect. 9.14], given by

$$c = \frac{\delta \tau}{1-\delta} {}_2F_1 (1, 1-\delta; 2-\delta; -\tau). \tag{2.63}$$

It follows from (2.62) that the coverage probability is independent of the BS density λ. This is an important property in single-antenna networks, called the *SIR invariance property*. It means that the coverage probability will not change when deploying more BSs. Intuitively, by deploying more BSs, the interference will increase. In the meantime, the minimal distance between the associated BS to the typical user r_0 also decreases, which will increase the signal power. Overall, the SIR distribution does not change.

The ASE, defined in (2.26), is then given by

$$R_{\mathrm{a}} = \lambda p_{\mathrm{c}} R_0 = \frac{\lambda}{1+c} \log_2(1+\tau). \tag{2.64}$$

Clearly, the ASE increases linearly with respect to the BS density λ. This is the foundation of the theory of "cell densification", which means the network capacity can be increased linearly by deploying more and more BSs. Furthermore, the energy efficiency of the network is written as

$$\xi_{\mathrm{EE}} = \frac{\log_2(1+\tau)}{(1+c)\left(\frac{1}{\xi}P_{\mathrm{t}} + P_{\mathrm{c}} + P_0\right)}, \tag{2.65}$$

which also indicates that increasing the BS density will not affect the energy efficiency.

Remark 2.1 Note that the SIR invariance property is obtained from the analysis of simple network models. As will be shown later in this monograph, e.g., in Sects. 4.1 and 5.2, this favorable property may not hold anymore for other models.

2.4.2 Ad Hoc Networks

In this subsection, we present the SINR distribution of ad hoc networks, as well as some common performance metrics. The difference from the cellular networks will be illustrated.

Consider a single-antenna ad hoc network with Rayleigh fading channels. The system model is described in Sect. 2.1.4. By substituting (2.50) into (2.55), and knowing $s = \tau r_0^{\alpha}$, the coverage probability is expressed as

$$p_{\mathrm{c}} = \exp\left\{-\tau r_0^{\alpha} \sigma_N^2 - \pi \lambda r_0^2 \tau^{\delta} \frac{\pi\delta}{\sin(\pi\delta)}\right\}. \tag{2.66}$$

From the above expression, we have two observations: (1) For a fixed density, decreasing the distance between the transmitter and receiver r increases the coverage probability; (2) Increasing the density of the transmitter–receiver pairs λ decreases the coverage probability. The first observation is intuitive because when the interference power does not change, making the receiver closer to its transmitter will enhance the transmit power, and therefore increase the coverage probability. The second observation is because deploying more transmitters increases the interference power, while the signal power does not change, given the fixed communication distance. Thus, the coverage probability decreases.

The ASE of the network is given by

$$R_{\mathrm{a}} = \lambda \exp\left\{-\tau r_0^{\alpha}\sigma_n^2 - \pi\lambda r_0^2\tau^{\delta}\frac{\pi\delta}{\sin(\pi\delta)}\right\}\log_2(1+\tau), \tag{2.67}$$

from which we can find that there is an optimal density $\lambda^{\star}$ that achieves the maximal ASE. We calculate the optimal density, given by

$$\lambda^{\star} = \frac{\sin(\pi\delta)}{\pi^2\delta r_0^2\tau}. \tag{2.68}$$

Intuitively, deploying more transmitter–receiver pairs directly increases the number of simultaneous communication links, but decreases the success probability of each link. Therefore, the ASE first increases and then decreases by increasing λ from 0 to infinity.

In ad hoc networks, there is another metric which is also commonly adopted. This metric is called the *transmission capacity (TC)*, which is defined as the maximum density of successful transmissions subject to certain p_{c} requirement [9]. Because the coverage probability decreases when the density increases, it is important to know the maximum density that the network can afford while the performance can be satisfied. Mathematically, for a given coverage probability requirement $1-\varepsilon$, the TC is given by

$$\mathrm{TC} = (1-\varepsilon)\lambda_{\varepsilon}, \tag{2.69}$$

where λ_{ε} is called the *optimal contention density*, and $p_{\mathrm{c}}(\lambda_{\varepsilon}) = 1-\varepsilon$ for $\varepsilon \in [0,1]$. Based on (2.66), the TC can be written as

$$\mathrm{TC} = \max\left(0, \frac{\ln(1-\varepsilon)^{-1} - \tau r_0^{\alpha}\sigma_N^2}{(1-\varepsilon)^{-1}\pi r_0^2\tau^{\delta}\frac{\pi\delta}{\sin(\pi\delta)}}\right). \tag{2.70}$$

It shows that if the noise power is large enough, it is impossible to satisfy the coverage requirement. Interested readers can find more analysis and results about the TC in [20].

Bibliographical Notes

Early history of analyzing Poisson network models has been introduced in Sect. 2.1.1. Besides, there are plenty of studies on Poisson network models in recent years. A series of works on the TC analysis of ad hoc networks using the PPP model appeared in 2000s. Specifically, the metric of TC was proposed by Weber et al. (2005), and then analyzed by Hunter et al. (2008), Huang et al. (2009), Ganti et al. (2011), etc.

The Poisson network models was also applied in the cognitive radio by Ghasemi et al. (2008), Rabbachin et al. (2011), Lee et al. (2012), ElSawy et al. (2013), etc., and in the device-to-device (D2D) networks by ElSawy et al. (2014), Sakr et al. (2015), Yang et al. (2016), etc. Single-tier cellular networks has been investigated via Poisson network models by Koutouris et al. (2012), Li et al. (2014), etc., while Poisson multitier networks has been studied by Chandrasekhar et al. (2009), Quek et al. (2011), Dhillon et al. (2012), etc.

References

1. E.N. Gilbert, Random plane networks. J. Soc. Ind. Appl. Math. **9**(4), 533–543 (1961)
2. L. Kleinrock, J. Silvester, Optimum transmission radii for packet radio networks or why six is a magic number, in Conference record: national telecommunications conference, (Birmingham, Alabama), pp. 4.3.1-4.3.5, Dec. 1978
3. L. Kleinrock, J. Silvester, Spatial reuse in multihop packet radio networks. Proc. IEEE **75**, 156–167 (1987)
4. H. Takagi, L. Kleinrock, Optimal transmission ranges for randomly distributed packet radio terminals. IEEE Trans. Commun. **32**, 246–257 (1984)
5. T.-C. Hou, V. Li, Transmission range control in multihop packet radio networks. IEEE Trans. Commun. **34**, 38–44 (1986)
6. F. Baccelli, B. Blaszczyszyn, P. Muhlethaler, Stochastic analysis of spatial and opportunistic ALOHA. IEEE J. Sel. Areas Commun. **27**, 1105–1119 (2009)
7. M.Z. Win, P.C. Pinto, L.A. Shepp, A mathematical theory of network interference and its applications. Proc. IEEE **97**, 205–230 (2009)
8. R.H.Y. Louie, M.R. McKay, I.B. Collings, Open-loop spatial multiplexing and diversity communications in ad hoc networks. IEEE Trans. Inf. Theor. **57**, 317–344 (2011)
9. S.P. Weber, X. Yang, J.G. Andrews, G. de Veciana, Transmission capacity of wireless ad hoc networks with outage constraints. IEEE Trans Inf. Theor. **51**, 4091–4102 (2005)
10. A.M. Hunter, J.G. Andrews, S. Weber, Transmission capacity of ad hoc networks with spatial diversity. IEEE Trans. Wirel. Commun. **7**, 5058–5071 (2008)
11. M. Kountouris, J.G. Andrews, Transmission capacity scaling of SDMA in wireless ad hoc networks, in Proceedings 2009 IEEE Information Theory Workshop, (Volos, Greece), pp. 534-538, Oct. 2009
12. M. Haenggi, J.G. Andrews, F. Baccelli, O. Dousse, M. Franceschetti, Stochastic geometry and random graphs for the analysis and design of wireless networks. IEEE J. Sel. Areas Commun. **27**, 1029–1046 (2009)
13. J.G. Andrews, R.K. Ganti, M. Haenggi, N. Jindal, S. Weber, A primer on spatial modeling and analysis in wireless networks. IEEE Commun. Mag. **48**, 156–163 (2010)
14. F. Baccelli, M. Klein, M. Lebourges, S. Zuyev, Stochastic geometry and architecture of communication networks. Telecommun. Syst. **7**, 209–227 (1997)
15. F. Baccelli, S. Zuyev, Stochastic geometry models of mobile communication networks, in Frontiers in queueing (CRC Press, 1997), pp. 227-243
16. J.G. Andrews, F. Baccelli, R.K. Ganti, A tractable approach to coverage and rate in cellular networks. IEEE Trans. Commun. **59**, 3122–3134 (2011)
17. M. Haenggi, R.K. Ganti, Interference in large wireless networks (vol. 3. Now Publishers Inc., 2009)
18. F. Baccelli, B. Błaszczyszyn, Stochastic geometry and wireless networks: volume I theory (vol. 3. Now Publishers Inc., 2009)
19. F. Baccelli, B. Błaszczyszyn, Stochastic geometry and wireless networks: volume II applications (vol. 4. Now Publishers Inc., 2009)

20. S. Weber, J.G. Andrews, Transmission capacity of wireless networks (vol. 5. Now Publishers Inc., 2012)
21. M. Haenggi, *Stochastic geometry for wireless networks* (Cambridge University Press, Cambridge, U.K., 2012)
22. S. Mukherjee, *Analytical modeling of heterogeneous cellular networks* (Cambridge University Press, 2014)
23. H. ElSawy, E. Hossain, M. Haenggi, Stochastic geometry for modeling, analysis, and design of multi-tier and cognitive cellular wireless networks: a survey. IEEE Commun. Surv. Tuts. **15**, 996–1019 (2013)
24. S.N. Chiu, D. Stoyan, W.S. Kendall, J. Mecke, *Stochastic geometry and its applications* (Wiley, 2013)
25. C. Li, J. Zhang, K.B. Letaief, Throughput and energy efficiency analysis of small cell networks with multi-antenna base stations. IEEE Trans. Wirel. Commun. **13**, 2505–2517 (2014)
26. N. Devroye, M. Vu, V. Tarokh, Cognitive radio networks. IEEE Signal Process. Mag. **25**, 12–23 (2008)
27. Z. Hasan, H. Boostanimehr, V.K. Bhargava, Green cellular networks: a survey, some research issues and challenges. IEEE Commun. Surv. Tuts. **13**, 524–540 (2011)
28. J. Xu, L. Qiu, Energy efficiency optimization for MIMO broadcast channels. IEEE Trans. Wirel. Commun. **12**, 690–701 (2013)
29. D. Nguyen, L.-N. Tran, P. Pirinen, M. Latva-aho, Precoding for full duplex multiuser MIMO system: spectral and energy efficiency maximization. IEEE Trans. Signal Process. **61**, 4038–4050 (2013)
30. T.D. Novlan, H.S. Dhillon, J.G. Andrews, Analytical modeling of uplink cellular networks. IEEE Trans. Wirel. Commun. **12**, 2669–2679 (2013)
31. Y. Wang, M. Haenggi, Z. Tan, The meta distribution of the SIR for cellular networks with power control. IEEE Trans. Commun. **66**, 1745–1757 (2018)
32. D. Zwillinger, *Table of integrals, series, and products* (Elsevier, Amsterdam, Netherlands, 2014)

Chapter 3
An Analytical Framework for Multi-Antenna Wireless Networks

Abstract This chapter presents a general analytical framework for large-scale multi-antenna wireless networks. We first introduce a general wireless network model, along with a brief survey of multi-antenna transmission techniques. Using tools from stochastic geometry, a unified framework is then presented for the tractable analysis of the multi-antenna wireless network model. To illustrate the effectiveness of this analytical framework, tractable expressions for the coverage analysis in both ad hoc and cellular networks are derived. It is shown that the presented framework makes the analysis of multi-antenna networks almost as tractable as single-antenna ones. Furthermore, it helps analytically gain key network design insights, such as revealing the impacts of the antenna size and network density.

3.1 Modeling Multi-Antenna Wireless Networks

3.1.1 Network Model and Notations

Consider the downlink transmission[1] of a multi-antenna wireless network, where each transmitter is equipped with N_t antennas, while N_r antennas are deployed at each receiver.[2] The transmitters are assumed to be spatially located according to a homogeneous PPP, denoted by Φ. Spatial division multiple access (SDMA), or multiuser MIMO, is adopted, and thus each transmitter may serve multiple users. Before we present details of the network model, general indexing rules are first given as below.

- For notations that are only related to a transmitter itself, we use the location of this transmitter for indexing. For example, the number of receivers that are communicated with the transmitter located at $x \in \Phi$ is denoted as U_x.

[1]In this monograph, we mainly focus on the performance analysis for downlink transmission while the uplink will be briefly discussed in Chap. 6.

[2]The numbers of antennas deployed at different transmitters and receivers can be different. However, they are set as two values, i.e., N_t and N_r for the transmitter and receiver, respectively, for the ease of presentation.

X. Yu et al., *Stochastic Geometry Analysis of Multi-Antenna Wireless Networks*, https://doi.org/10.1007/978-981-13-5880-7_3

- For notations only related to a single receiver, we use two letters in the subscript. The former represents the index of the receiver, while the latter indicates the location of the transmitter with which this receiver is associated. For example, the number of data streams transmitted from the transmitter located at x to its u-th receiver ($0 \leq u \leq U_x - 1$) is denoted as D_{ux}.
- For notations that are related to a pair of transmitter and receiver, we use three letters in the subscript to index. The first letter represents the index of the receiver, the second one indicates the location of the transmitter with which the receiver is associated, and the last letter displays the location of the transmitter in the pair. For instance, the channel between the transmitter at x and the u-th receiver associated with the transmitter located at x' is denoted as $\mathbf{H}_{ux'x}$. As a special case, when the receiver is associated with and served by the considered transmitter, the second and third indices are the same.

We assume that a transmitter located at x communicates with U_x receivers. For the u-th receiver, $D_{ux} \leq \min\{N_\mathrm{t}, N_\mathrm{r}\}$ independent data streams are assumed to be sent by the corresponding transmitter, where $0 \leq u \leq U_x - 1$. The power allocation matrix for the u-th receiver is denoted as $\mathbf{P}_{ux} = \mathrm{diag}\left(P_{ux,1}, P_{ux,2}, \ldots, P_{ux,D_{ux}}\right)$, where $P_{ux,d}$ is the power allocated to the d-th data stream transmitted to the u-th user served by the transmitter located at x. We focus on the performance analysis of the typical receiver at the origin (the 0-th receiver), and the location of its associated transmitter is denoted as x_0. There will be *intracell interference*, from signals for other receivers served by the same transmitter, as well as *intercell interference* from other transmitters. In this way, the received $D_{0x_0} \times 1$ data vector at the typical receiver is given by

$$\begin{aligned}\hat{\mathbf{s}}_{0x_0} &= \underbrace{r_0^{-\frac{\alpha}{2}} \mathbf{W}_{0x_0}^H \mathbf{H}_{0x_0x_0} \mathbf{F}_{0x_0} \sqrt{\mathbf{P}_{0x_0}} \mathbf{s}_{0x_0}}_{\text{desired information signal}} + \underbrace{r_0^{-\frac{\alpha}{2}} \mathbf{W}_{0x_0}^H \mathbf{H}_{0x_0x_0} \sum_{u=1}^{U_{x_0}-1} \mathbf{F}_{ux_0} \sqrt{\mathbf{P}_{ux_0}} \mathbf{s}_{ux_0}}_{\text{intracell interference}} \\ &\quad + \underbrace{\mathbf{W}_{0x_0}^H \sum_{x \in \Phi'} \|x\|^{-\frac{\alpha}{2}} \mathbf{H}_{0x_0x} \sum_{u=0}^{U_x-1} \mathbf{F}_{ux} \sqrt{\mathbf{P}_{ux}} \mathbf{s}_{ux}}_{\text{intercell interference}} + \underbrace{\mathbf{W}_{0x_0}^H \mathbf{n}_{0x_0}}_{\text{noise}}.\end{aligned} \tag{3.1}$$

There are many notations in (3.1), which shall be commonly used in this chapter. We introduce these notations in the following four parts.

Propagation and Noise

The transmitted data streams from the transmitter located at x to its u-th associated receiver are denoted as $\mathbf{s}_{ux} \in \mathbb{C}^{D_{ux} \times 1}$ such that $\mathbb{E}\left[\mathbf{s}_{ux}\mathbf{s}_{ux}^H\right] = \mathbf{I}_{D_{ux}}$, and $\mathbf{n}_{ux} \sim \mathcal{CN}(\mathbf{0}_{N_\mathrm{r}}, N_0 \mathbf{I}_{N_\mathrm{r}})$ is the complex additive white Gaussian noise vector at the u-th user associated with the transmitter at x.

For the large-scale path loss, we adopt the standard power law path loss model in (2.17). The value of the path loss exponent α is normally in the range between 2 and 4, where 2 is for propagation in free space. $\alpha = 4$ is for relatively lossy environments or for the case of full specular reflection from the earth surface—the so-called flat earth model, and it is the typical value used in many works to simplify the analytical results. In some environments, such as buildings, stadiums, and other indoor environments, the path loss exponent can reach values in the range of 4 to 6. In this chapter, α assumes any value larger than 2. Throughout the monograph, we often adopt the notation $\delta = 2/\alpha$ for the neatness of presentation.

The matrix $\mathbf{H}_{ux'x} \in \mathbb{C}^{N_r \times N_t}$ denotes the small-scale fading from the transmitter located at x to the u-th receiver associated with the transmitter at x', whose coefficients are random variables distributed according to different fading channels, such as Rayleigh or Nakagami fading channels.

Precoder and Combiner

The transmitter at location x adopts a precoder for transmitting to its u-th associated receiver, denoted as $\mathbf{F}_{ux} \in \mathbb{C}^{N_t \times D_{ux}}$, while the combiner used by the u-th receiver tagged with the transmitter at x is denoted as $\mathbf{W}_{ux} \in \mathbb{C}^{N_r \times D_{ux}}$. Both the precoder and combiner depend on the adopted MIMO transmission schemes and will be discussed in detail in the next section. The precoders and combiners can be designed according to different levels of CSI availability. In particular, they can be designed with the instantaneous CSI, i.e., accurate information of $\mathbf{H}_{ux'x}$ at each time instant, or based on statistical CSI with some long-term information of $\mathbf{H}_{ux'x}$, such as the channel covariance matrix.

Desired Information Signal

The received power of the desired information signal critically determines the communication performance, which is given by

$$S = r_0^{-\alpha} \left\| \mathbf{W}_{0x_0}^H \mathbf{H}_{0x_0x_0} \mathbf{F}_{0x_0} \sqrt{\mathbf{P}_{0x_0}} \mathbf{s}_{0x_0} \right\|_2^2 . \tag{3.2}$$

In particular, in addition to the power gain provided by the adopted MIMO techniques and the transmit power, it also depends on the distance from the typical receiver to its associated transmitter located at x_0, denoted as r_0, i.e., $r_0 \triangleq \|x_0\|$. The value of x_0 depends on the adopted transmitter–receiver association strategy, which is typically based on the average received power, e.g., bipolar association in ad hoc networks [1], the nearest-transmitter association in cellular networks [2], and biased association in HetNets [3]. According to different strategies, r_0 is either a deterministic value or a random variable with the probability density function (pdf) denoted as $f_{r_0}(r)$. Specific examples shall be presented later to show the effect of cell association strategies on the distance r_0.

Interference

In wireless networks, interference typically is the limiting factor for performance. Since the transmitter associated with the typical receiver is located at x_0, the set of interfering transmitters is naturally given by $\Phi \backslash \{x_0\}$. Nevertheless, we do not use this notation in (3.1). Instead, we adopt Φ' to denote the equivalent point process of the interfering transmitters, which helps clarify the spatial distribution of the interferers and help derive the Laplace transform in later sections. The point process Φ' in general can be a union of different point processes consisting of J $(J \geq 1)$ types of interferers, i.e., $\Phi' = \cup_{j=1}^{J} \Phi'_j$. In particular, we assume that the interferers belonging to the j-th type are distributed according to a PPP Φ'_j conditioned on the disk $b(o, l_j(r_0))$ being empty.

The radius $l_j(r_0)$ of the disk is a key parameter in network performance analysis. It is the minimum distance between the typical receiver and the *interfering transmitters* of the j-th type, and the functions $l_j(\cdot)$ are critically determined by the cell association strategy. Once the functions $l_j(\cdot)$ are specified, the point process Φ' that describes the set of interfering transmitters is then accordingly determined.

This model not only reflects the general multi-tier HetNet setting but also applies when the interferers have different channel power gain distributions in a single-tier network. In the following, we give some examples to illustrate how specific network settings and cell association strategies determine the association distance r_0, the key parameter $l_j(r_0)$, and the corresponding relation between Φ' and Φ.

Example 3.1 (Ad hoc networks) As introduced in Sect. 2.1.4, in ad hoc networks, a Poisson bipolar model is adopted. In particular, the typical receiver is associated with a transmitter located at a deterministic distance r_0. According to Slivnyak's theorem, the distribution of $\Phi \backslash \{x_0\}$ is the same as Φ. Therefore, there is no loss to set $J = 1$, and the equivalent point process of the interfering nodes $\Phi' = \Phi$, i.e., $l(r_0) = 0$ in the ad hoc network analysis, which does not affect the distribution of the aggregated interference.

Example 3.2 (Cellular networks with different types of transmitters) Consider a cellular network with J types of BSs, i.e., $\Phi = \cup_{j=1}^{J} \Phi_j$, where each type of BSs is spatially distributed as a homogeneous PPP Φ_j with the density λ_j. We assume that the typical user is associated with the nearest BS of the k-th type. Therefore, according to (2.60), we have the distribution of r_0 as

$$f_{r_0}(r) = 2\pi \lambda_k r e^{-\pi \lambda_k r^2}, \tag{3.3}$$

and the minimum distance between the typical user and the interfering BSs of the j-th type is given by

$$l_j(r_0) = \begin{cases} r_0 & j = k \\ 0 & j \neq k. \end{cases} \tag{3.4}$$

In other words, the equivalent point process $\Phi' = \cup_{j=1}^{J} \Phi'_j$ is determined as follows:

- For $j \neq k$, the equivalent point process is $\Phi'_j = \Phi_j$.
- For $j = k$, Φ'_k is a PPP with density λ_k conditioned on the disk $b(o, r_0)$ being empty.

Example 3.3 (K-tier heterogeneous cellular networks) In heterogeneous cellular networks with K tiers, i.e., $\Phi = \cup_{j=1}^{K} \Phi_j$, where the BSs in each tier of the cellular network form a PPP with the density λ_j, a commonly adopted cell association rule is that the user is associated with the k-th tier if

$$k = \arg \max_{j \in \{1,\dots,K\}} P_j B_j r_j^{-\alpha}, \tag{3.5}$$

where P_j is the transmit power of the BSs in the j-th tier, and r_j is the distance between the typical user and its nearest BS in the j-th tier. In HetNets, mobile users need to be actively pushed onto the more lightly loaded tiers to optimize the overall network performance, which is reflected in the load balancing bias factor B_j of the j-th tier [4]. In this way, conditioned on the typical user being associated with the k-th tier, i.e., $r_0 = r_k$, the distribution of r_k is given by [4, Lemma 3]

$$f_{r_k}(r) = 2\pi r \left[\sum_{j=1}^{K} \lambda_j \left(\frac{P_j B_j}{P_k B_k} \right)^{\frac{2}{\alpha}} \right] e^{-\pi r^2 \left[\sum_{j=1}^{K} \lambda_j \left(\frac{P_j B_j}{P_k B_k} \right)^{\frac{2}{\alpha}} \right]}, \tag{3.6}$$

and the minimum distance between the typical user and the BS in the j-th tier is given by

$$l_j(r_k) = \left(\frac{P_j B_j}{P_k B_k} \right)^{\frac{1}{\alpha}} r_k. \tag{3.7}$$

In this way, each equivalent point process Φ'_j of the j-th tier is a PPP with density λ_j conditioned on the disk $b\left(o, \left(\frac{P_j B_j}{P_k B_k}\right)^{\frac{1}{\alpha}} r_k\right)$ to be empty.

3.1.2 *Signal-to-Interference-Plus-Noise Ratio (SINR)*

As discussed in Sect. 2.4, the receive SINR determines the communication performance. Next, we derive the SINR expression for general multiuser MIMO networks. Before the derivation, we introduce two notations that help the presentation. We denote the d-th column of the precoder matrix $\mathbf{F}_{ux}$ and combiner matrix $\mathbf{W}_{ux}$ as $\mathbf{f}_{ux,d}$ and $\mathbf{w}_{ux,d}$, respectively, which are the precoding and combining vectors for the d-th data stream transmitted to the u-th receiver associated with the transmitter located at x.

According to (3.1), the power of the desired signal for the d-th data stream of the typical receiver is given by

$$\begin{aligned} S_{0x_0,d} &= P_{0x_0,d} r_0^{-\alpha} \mathbf{w}_{0x_0,d}^H \mathbf{H}_{0x_0x_0} \mathbf{f}_{0x_0,d} \mathbf{f}_{0x_0,d}^H \mathbf{H}_{0x_0x_0}^H \mathbf{w}_{0x_0,d} \\ &= r_0^{-\alpha} \left| \left[\mathbf{W}_{0x_0}^H \mathbf{H}_{0x_0x_0} \mathbf{F}_{0x_0} \sqrt{\mathbf{P}_{0x_0}} \right]_d \right|^2, \end{aligned} \tag{3.8}$$

where $[\cdot]_d$ represents the d-th diagonal element of a square matrix. In addition, the power of the inter-stream interference from the transmitter at x_0 can be written as

$$\begin{aligned} I_{0x_0,d}^{\text{inter-stream}} &= r_0^{-\alpha} \mathbf{w}_{0x_0,d}^H \mathbf{H}_{0x_0x_0} \left(\sum_{i \neq d} \mathbf{f}_{0x_0,i} P_{0x_0,i} \mathbf{f}_{0x_0,i}^H \right) \mathbf{H}_{0x_0x_0}^H \mathbf{w}_{0x_0,d} \\ &= r_0^{-\alpha} \left[\mathbf{W}_{0x_0}^H \mathbf{H}_{0x_0x_0} \mathbf{F}_{0x_0 \backslash d} \mathbf{P}_{0x_0 \backslash d} \mathbf{F}_{0x_0 \backslash d}^H \mathbf{H}_{0x_0x_0}^H \mathbf{W}_{0x_0} \right]_d , \end{aligned} \tag{3.9}$$

where $\mathbf{F}_{0x_0 \backslash d}$ consists of all the columns in $\mathbf{F}_{0x_0}$ except the d-th column, and $\mathbf{P}_{0x_0 \backslash d} = \text{diag}\left(P_{0x_0,0}, P_{0x_0,1}, \ldots, P_{0x_0,d-1}, P_{0x_0,d+1}, \ldots, P_{0x_0,D_{0x_0}-1} \right)$. Similarly, the power of the intracell interference is given by

$$\begin{aligned} I_{0x_0,d}^{x_0} &= r_0^{-\alpha} \left[\mathbf{W}_{0x_0}^H \mathbf{H}_{0x_0x} \left(\sum_{u=1}^{U_{x_0}-1} \sum_{i=0}^{D_{ux_0}} \mathbf{f}_{ux_0,i} P_{ux_0,i} \mathbf{f}_{ux_0,i}^H \right) \mathbf{H}_{0x_0x}^H \mathbf{W}_{0x_0} \right]_d \\ &= r_0^{-\alpha} \left[\mathbf{W}_{0x_0}^H \mathbf{H}_{0x_0x} \mathbf{F}_{x_0 \backslash 0} \mathbf{P}_{x_0 \backslash 0} \mathbf{F}_{x_0 \backslash 0}^H \mathbf{H}_{0x_0x}^H \mathbf{W}_{0x_0} \right]_d , \end{aligned} \tag{3.10}$$

where

$$\begin{aligned} \mathbf{P}_{x_0 \backslash 0} &= \text{diag}\left(\mathbf{P}_{1x_0}, \mathbf{P}_{2x_0}, \ldots, \mathbf{P}_{(U_{x_0}-1)x_0} \right) \\ \mathbf{F}_{x_0 \backslash 0} &= \left[\mathbf{F}_{1x_0}, \mathbf{F}_{2x_0}, \ldots, \mathbf{F}_{(U_{x_0}-1)x_0} \right] \end{aligned} \tag{3.11}$$

are the composite matrices for all the receivers served by the transmitter at x_0, except the typical receiver. Furthermore, the power gain for the intercell interference from the transmitter located at x is given by

$$I_{0x_0,d}^{x} = \|x\|^{-\alpha} \left[\mathbf{W}_{0x_0}^H \mathbf{H}_{0x_0x} \mathbf{F}_x \mathbf{P}_x \mathbf{F}_x^H \mathbf{H}_{0x_0x}^H \mathbf{W}_{0x_0} \right]_d , \tag{3.12}$$

where

$$\begin{aligned} \mathbf{F}_x &= \left[\mathbf{F}_{0x}, \mathbf{F}_{1x}, \ldots, \mathbf{F}_{(U_x-1)x} \right] \\ \mathbf{P}_x &= \text{diag}\left(\mathbf{P}_{0x}, \mathbf{P}_{1x}, \ldots, \mathbf{P}_{(U_x-1)x} \right) . \end{aligned} \tag{3.13}$$

Therefore, the SINR value for the d-th data stream of the typical receiver in multi-antenna systems is written by

$$\begin{aligned} &\text{SINR}_{0x_0,d} \\ &= \frac{S_{0x_0,d}}{N_0 \left[\mathbf{W}_{0x_0}^H \mathbf{W}_{0x_0} \right]_d + \left(I_{0x_0,d}^{\text{inter-stream}} + I_{0x_0,d}^{x_0} \right) + \sum_{x \in \Phi'} I_{0x_0,d}^{x}} \\ &= \frac{r_0^{-\alpha} \left| \left[\mathbf{W}_{0x_0}^H \mathbf{H}_{0x_0x_0} \mathbf{F}_{0x_0} \sqrt{\mathbf{P}_{0x_0}} \right]_d \right|^2}{N_0 \left[\mathbf{W}_{0x_0}^H \mathbf{W}_{0x_0} \right]_d + \sum_{x \in \Phi} \|x\|^{-\alpha} \left[\mathbf{W}_{0x_0}^H \mathbf{H}_{0x_0x} \mathbf{F}_x \mathbf{P}_x \mathbf{F}_x^H \mathbf{H}_{0x_0x}^H \mathbf{W}_{0x_0} \right]_d} , \end{aligned} \tag{3.14}$$

where

$$\mathbf{F}_x = \begin{cases} \left[\mathbf{F}_{0x\backslash d}, \mathbf{F}_{1x}, \mathbf{F}_{2x}, \ldots, \mathbf{F}_{(U_x-1)x}\right] & x = x_0, \\ \left[\mathbf{F}_{0x}, \mathbf{F}_{1x}, \ldots, \mathbf{F}_{(U_x-1)x}\right] & \text{otherwiese}, \end{cases}$$
$$\mathbf{P}_x = \begin{cases} \text{diag}\left(\mathbf{P}_{0x\backslash d}, \mathbf{P}_{1x}, \mathbf{P}_{2x}, \ldots, \mathbf{P}_{(U_x-1)x}\right) & x = x_0, \\ \text{diag}\left(\mathbf{P}_{0x}, \mathbf{P}_{1x}, \ldots, \mathbf{P}_{(U_x-1)x}\right) & \text{otherwiese} \end{cases} \tag{3.15}$$

are the composite power allocation and precoding matrices adopted by the transmitter at x. Note that when $x = x_0$, the power allocation and precoding vectors for the d-th data stream transmitted to the typical receiver should be deducted from the interference terms, which is different from the interference from the transmitters located at the other locations.

We would like to denote some components in (3.14) for the ease of presentation and illustration. Recall that in single-antenna wireless networks, the signal and interferer's power gains are typically determined by a scalar channel gain. In contrast, in multi-antenna wireless networks, the signal and interferer's power gains are critically determined by the adopted MIMO transmission schemes, i.e., the precoding matrices $\mathbf{F}_x$ and combining matrices $\mathbf{W}_{ux}$. Therefore, we define the normalized power gain of the desired signal as

$$g_{x_0,d} \triangleq \frac{\left|\left[\mathbf{W}_{0x_0}^H \mathbf{H}_{0x_0x_0} \mathbf{F}_{0x_0} \sqrt{\mathbf{P}_{0x_0}}\right]_d\right|^2}{\left[\mathbf{W}_{0x_0}^H \mathbf{W}_{0x_0}\right]_d}, \tag{3.16}$$

while the normalized interferer's power gain from the interfering transmitter located at x is defined as

$$g_{x,d} \triangleq \frac{\left[\mathbf{W}_{0x_0}^H \mathbf{H}_{0x_0x} \mathbf{F}_x \mathbf{P}_x \mathbf{F}_x^H \mathbf{H}_{0x_0x}^H \mathbf{W}_{0x_0}\right]_d}{\left[\mathbf{W}_{0x_0}^H \mathbf{W}_{0x_0}\right]_d}. \tag{3.17}$$

Different channel distributions and MIMO techniques lead to different distributions for $g_{x_0,d}$ and $g_{x,d}$. In the next section, several commonly adopted MIMO transmission techniques shall be introduced, and specific expressions of the power gains are correspondingly given.

Therefore, the SINR value is represented as

$$\text{SINR}_{0x_0,d} \triangleq \frac{g_{x_0,d} r_0^{-\alpha}}{\sigma_{\text{n}}^2 + \sum_{x \in \Phi'} g_{x,d} \|x\|^{-\alpha}}. \tag{3.18}$$

Following the previous derivation, σ_{n}^2 in (3.18) should be equal to the noise power N_0. We introduce this new notation σ_{n}^2 to denote the normalized noise power. It will be used in the following chapters and equals N_0 normalized by some common terms in both the signal and interference, where the normalization is to keep expression (3.18) clean.

3.2 Multi-Antenna Techniques

Multi-antenna, multiple-input multiple-output (MIMO), techniques have drawn much attention from both academia and industry, and they are a key enabler for broadband wireless communication systems, such as 4G LTE networks. In the physical layer design of wireless systems, there are two vital gains provided by multiple antennas. The first one is called the *spatial diversity gain*. Different from the diversity gains in the time domain and/or frequency domain in conventional communications systems, the diversity gains by deploying multiple antennas are obtained from the spatial domain, without expanding the bandwidth or increasing the transmit power. Specifically, the existence of multiple antennas in a system means there exist multiple independent propagation paths. This approach mainly aims at improving the reliability of communications by transmitting the same data stream via different spatial paths. The second gain is called the *spatial multiplexing gain*, obtained by transmitting multiple data streams simultaneously through different spatial paths in parallel. To achieve this kind of multiplexing gains, delicate design of MIMO transmission schemes are needed, involving both the precoder and combiner.

When it comes to the network layer design of wireless networks, the MIMO techniques are mainly used for mitigating the co-channel interference (CCI) [5, 6]. A general approach is to perform precoding/beamforming and decoding/combining at transmitter and receiver sides, respectively, to mitigate CCI while keeping the signal power to the intended receivers at a desirable level [7]. In this section, we introduce several commonly adopted MIMO transmission techniques in different MIMO system settings, and the correspondingly signal and interference distributions are also presented.

3.2.1 Single-User MIMO

We start with a simple network setting, where each transmitter only communicates with one receiver at a time ($U = 1$), and both sides employ multiple antennas, i.e., single-user MIMO. This corresponds to the case that the wireless network operates in a time division multiple access (TDMA) fashion. In this mode, the focus is primarily on the desired signal link, while neglecting the interference from other transmitters. The multiple antennas are mainly used for canceling inter-stream interference and/or maximizing the desired signal strength.

A simple yet well-known MIMO processing technique for single-stream transmission is called the maximum ratio transmit (MRT). For each transmitter at x, it adopts the beamforming vector

$$\mathbf{f}_x = \frac{\mathbf{H}_{xx}^H \mathbf{w}_x}{\|\mathbf{H}_{xx}^H \mathbf{w}_x\|_2}, \tag{3.19}$$

where $\mathbf{w}_x$ is the combining vector that will be determined in the following. In particular, the MRT beamforming vector is the normalized effective channel $\mathbf{H}_{xx}^H \mathbf{w}_x$. In this way, the power of the desired signal is given by

$$\sqrt{P_x}\|x\|^{-\frac{\alpha}{2}} \mathbf{w}_x^H \mathbf{H}_{xx} \mathbf{f}_x s_x = \sqrt{P_x}\|x\|^{-\frac{\alpha}{2}} \|\mathbf{H}_{xx}^H \mathbf{w}_x\|_2 s_x, \tag{3.20}$$

and the resulting SNR is

$$\mathrm{SNR} = \frac{P_x \|x\|^{-\alpha} \|\mathbf{H}_{xx}^H \mathbf{w}_x\|_2^2}{\|\mathbf{w}_x\|_2^2}. \tag{3.21}$$

To maximize the SNR, the combining vector $\mathbf{w}_x$ has to be the dominant *eigenvector* of $\mathbf{H}_{xx}\mathbf{H}_{xx}^H$, denoted by $\vec{\lambda}_{\max}\left(\mathbf{H}_{x_0x_0}\mathbf{H}_{x_0x_0}^H\right)$. Therefore, the normalized signal channel power gain with MRT is

$$g_{x_0} = \frac{\left\|\mathbf{H}_{x_0x_0}^H \vec{\lambda}_{\max}\left(\mathbf{H}_{x_0x_0}\mathbf{H}_{x_0x_0}^H\right)\right\|_2^2}{\left\|\vec{\lambda}_{\max}\left(\mathbf{H}_{x_0x_0}\mathbf{H}_{x_0x_0}^H\right)\right\|_2^2} = P_{x_0}\lambda_{\max}\left(\mathbf{H}_{x_0}\mathbf{H}_{x_0}^H\right), \tag{3.22}$$

where $\lambda_{\max}$ denotes the largest *eigenvalue* of a square matrix. The normalized interferer's channel power gain g_x is expressed by

$$g_x = \frac{P_x\left[\vec{\lambda}_{\max}^H\left(\mathbf{H}_{x_0x_0}\mathbf{H}_{x_0x_0}^H\right)\mathbf{H}_{x_0x}\mathbf{H}_{xx}^H\vec{\lambda}_{\max}\left(\mathbf{H}_{xx}\mathbf{H}_{xx}^H\right)\right]^2}{\left\|\mathbf{H}_{xx}^H\vec{\lambda}_{\max}\left(\mathbf{H}_{xx}\mathbf{H}_{xx}^H\right)\right\|_2^2}. \tag{3.23}$$

The main limitation of MRT is that it only supports single-stream transmission, and therefore the multiple antennas only provide array gains. As for multi-stream transmission, singular vector decomposition (SVD) based transmission is one of the most popular MIMO schemes that can provide multiplexing gains. In MIMO-SVD transmission, the precoding and combining matrices are given by

$$\mathbf{F}_{xx} = \mathbf{V}_{xx}(:, [1:D]), \quad \mathbf{W}_{xx} = \mathbf{U}_{xx}(:, [1:D]), \tag{3.24}$$

respectively, where $\mathbf{H}_{xx} = \mathbf{U}_{xx}\Lambda_{xx}\mathbf{V}_{xx}^H$ is the SVD of the channel matrix $\mathbf{H}_{xx}$. In this way, the physical channel is decoupled into D parallel paths without inter-stream interference during the transmission, and thus the detection at the receiver side is simplified. Correspondingly, the gains for the signal channel power and interferer's power for MIMO-SVD are given by

$$\begin{aligned} g_{x_0,d} &= \sigma_{x_0,d}^2 P_{x_0,d}, \\ g_{x,d} &= \left[\mathbf{U}_{x_0x_0}^H(:, [1:D])\mathbf{H}_{x_0x}\mathbf{V}_{xx}(:, [1:D])\mathbf{P}_{xx}\mathbf{V}_{xx}^H(:, [1:D])\mathbf{H}_{x_0x}^H\mathbf{U}_{x_0x_0}(:, [1:D])\right]_d, \end{aligned} \tag{3.25}$$

where $\sigma_{x_0,d}$ is the d-th largest singular value of the channel matrix $\mathbf{H}_{x_0x_0}$.

Both MRT and MIMO-SVD belong to closed-loop MIMO, i.e., they require explicit feedback of channel state information from the receiver to the transmitter. In situations where such feedback is not available, the burden is shifted to the receiver side, which needs to perform vector detection to handle inter-stream interference. The zero-forcing (ZF) detector is a low-complexity option, where $D = N_\mathrm{t}$ data streams are transmitted and the decoding matrix $\mathbf{W}_x$ is designed as

$$\mathbf{W}_x^H = \mathbf{H}_{xx}^\dagger = \left(\mathbf{H}_{xx}^H \mathbf{H}_{xx}\right)^{-1} \mathbf{H}_{xx}^H. \tag{3.26}$$

Then each data stream is decoded separately. Hence, the signal power gain and interferer's power gain are written by

$$\begin{aligned} g_{x_0,d} &= \frac{P_{x_0,d}}{\left[(\mathbf{H}_{x_0x_0}^H \mathbf{H}_{x_0x_0})^{-1}\right]_d}, \\ g_{x,d} &= \frac{\left[\mathbf{H}_{x_0x_0}^\dagger \mathbf{H}_{x_0x} \mathbf{P}_x \mathbf{H}_{x_0x}^H \left(\mathbf{H}_{x_0x_0}^\dagger\right)^H\right]_d}{\left[(\mathbf{H}_{x_0x_0}^H \mathbf{H}_{x_0x_0})^{-1}\right]_d}. \end{aligned} \tag{3.27}$$

Note that the ZF detector only applies when $N_\mathrm{r} \geq N_\mathrm{t}$, as the matrix $\mathbf{H}_{xx}^H \mathbf{H}_{xx}$ is invertible only when $\mathbf{H}_{xx}$ is full rank in column. Therefore, zero-forcing with transmit antenna selection (ZF-TAS) is often performed when there are more antennas at the transmitter side than the receiver side.

Assume $L \leq N_\mathrm{r} < N_\mathrm{t}$ transmit antennas of each transmitter are selected to communicate with the receivers, and $D = L$ data streams are transmitted. The precoding and decoding matrices of the ZF-TAS scheme are given by

$$\mathbf{F}_x = \mathbf{S}_x, \quad \mathbf{W}_x^H = (\mathbf{H}_{xx}\mathbf{S}_x)^\dagger, \tag{3.28}$$

where $\mathbf{S}_x$ is an $N_\mathrm{t} \times L$ selection matrix formed by choosing L columns from $\mathbf{I}_{N_\mathrm{t}}$. With TAS, the power gains are correspondingly given by

$$\begin{aligned} g_{x_0,d} &= \frac{P_{x_0,d}}{\left[(\mathbf{S}_{x_0}^H \mathbf{H}_{x_0x_0}^H \mathbf{H}_{x_0x_0} \mathbf{S}_{x_0})^{-1}\right]_d}, \\ g_{x,d} &= \frac{\left[\left(\mathbf{H}_{x_0x_0}\mathbf{S}_{x_0}\right)^\dagger \mathbf{H}_{x_0x} \mathbf{S}_x \mathbf{P}_x \mathbf{S}_x^H \mathbf{H}_{x_0x}^H \left(\left(\mathbf{H}_{x_0x_0}\mathbf{S}_{x_0}\right)^\dagger\right)^H\right]_d}{\left[(\mathbf{S}_{x_0}^H \mathbf{H}_{x_0x_0}^H \mathbf{H}_{x_0x_0} \mathbf{S}_{x_0})^{-1}\right]_d}. \end{aligned} \tag{3.29}$$

Ideally, the selected transmit antennas should be chosen to maximize the SNR, i.e.,

$$\mathbf{S}_x^\star = \arg\max_{\mathbf{S}_x} \min_{d \in \{1,\ldots,D\}} \frac{1}{\left[\left(\mathbf{S}_x^H \mathbf{H}_{xx}^H \mathbf{H}_{xx} \mathbf{S}_x\right)^{-1}\right]_d}. \tag{3.30}$$

Nevertheless, the open-loop counterpart of this scheme is often considered due to its simplicity, for which the antennas are selected randomly [8].

Similarly, the ZF precoder can be directly applied when $N_t \geq N_r$, and zero-forcing with receive antenna selection (ZF-RAS) can be another choice when $N_t < N_r$, which are not explained in detail here for the simplicity of presentation.

3.2.2 *Multiuser MIMO*

For multiuser MIMO (MU-MIMO) systems [9], the capacity of the MIMO broadcasting channel, e.g., for downlink transmission in the cellular network, is achieved by the nonlinear dirty paper coding (DPC) [10], which is highly complex to implement in practice. Suboptimal approximations of DPC, including Tomlinson–Harashima precoding (THP), were then proposed to approach the capacity [11]. However, these precoding schemes are seldom adopted in practical systems due to their high complicity of implementation. Furthermore, the performance analysis with these complicated MIMO techniques is almost intractable even for the link-level analysis, i.e., point-to-point performance characterization. For practical use and tractable analysis, simpler linear precoding schemes are favored.

The ZF precoding can be similarly applied to MU-MIMO. In this part, we start from a simple case where each receiver is deployed with a single antenna, i.e., $N_r = 1$. In particular, for the transmitter located at x with its associated U_x single-antenna receivers, we have the composite precoding matrix and channel matrix as follows:

$$\begin{aligned} \mathbf{F}_x &= \left[\mathbf{f}_{0x}, \mathbf{f}_{1x}, \ldots, \mathbf{f}_{(U_x-1)x}\right], \\ \mathbf{H}_{xx} &= \left[\mathbf{h}_{0xx}^T, \mathbf{h}_{1xx}^T, \ldots, \mathbf{h}_{(U_x-1)xx}^T\right]^T. \end{aligned} \tag{3.31}$$

Therefore, the ZF precoding matrix is given by

$$\mathbf{F}_x = \mathbf{H}_{xx}^{\dagger} = \mathbf{H}_{xx}^H \left(\mathbf{H}_{xx}\mathbf{H}_{xx}^H\right)^{-1}. \tag{3.32}$$

Note that the condition that the ZF precoding can be realized is $N_t \geq U_x N_r = U_x$, which is the common case in the MU-MIMO scenario. Correspondingly, the signal power gain and interferer's power gain are given by

$$\begin{aligned} g_{x_0} &= \frac{P_{0x_0}}{w_{0x_0}^2}, \\ g_x &= \mathbf{h}_{0x_0x}^T \mathbf{F}_x \mathbf{P}_x \mathbf{F}_x^H \mathbf{h}_{0x_0x} / w_{0x_0}^2. \end{aligned} \tag{3.33}$$

On the other hand, when each of the U receivers has multiple antennas, i.e., $N_r > 1$, ZF precoder is fairly suboptimal since each receiver is able to coordinate the processing of its own received outputs. Similar to (3.31), if we define the composite channel matrix and precoding matrix as

$$\mathbf{F}_x = \left[\mathbf{F}_{0x}, \mathbf{F}_{1x}, \ldots, \mathbf{F}_{U-1x}\right], \\ \mathbf{H}_{xx} = \left[\mathbf{H}_{0xx}^T, \mathbf{H}_{1xx}^T, \ldots, \mathbf{H}_{U-1xx}^T\right]^T, \tag{3.34}$$

the optimal solution under the constraint that all interference between the receivers is canceled is achieved when $\mathbf{H}_{xx}\mathbf{F}_x$ is block diagonal. Hence, this approach is named as block diagonalization (BD) [12]. In other words, we should first find $\{\mathbf{M}_{uxx}\}_{u=0}^{U_x-1}$ to satisfy the condition that

$$\mathbf{H}_{u'xx}\mathbf{M}_{uxx} = \mathbf{0}, \quad \forall u' \neq u. \tag{3.35}$$

If we define a composite matrix for the u-th receiver associated with the transmitter located at x as

$$\tilde{\mathbf{H}}_{xx\backslash u} = \left[\mathbf{H}_{0xx}^T, \mathbf{H}_{1xx}^T, \ldots, \mathbf{H}_{u-1xx}^T, \mathbf{H}_{u+1xx}^T, \ldots \mathbf{H}_{U-1xx}^T\right]^T, \tag{3.36}$$

and compute its SVD as

$$\tilde{\mathbf{H}}_{xx\backslash u} = \tilde{\mathbf{U}}_{xx\backslash u} \begin{bmatrix} \tilde{\Lambda}_{xx\backslash u} & \mathbf{0} \\ \mathbf{0} & \mathbf{0} \end{bmatrix} \left[\tilde{\mathbf{V}}_{xx\backslash u}^1 \; \tilde{\mathbf{V}}_{xx\backslash u}^0\right]. \tag{3.37}$$

In this way, the matrix $\mathbf{M}_{uxx}$ must lie in the null space of $\tilde{\mathbf{H}}_{xx\backslash u}$ if all the interference between receivers needs to be canceled. Therefore, there is room for D-stream transmission to the u-th receiver if $N_t \geq \text{rank}(\tilde{\mathbf{H}}_{xx\backslash u}) - D$. Overall, the condition that BD can be achieved is that $N_t \geq \max\left\{\text{rank}(\tilde{\mathbf{H}}_{xx\backslash 1}), \text{rank}(\tilde{\mathbf{H}}_{xx\backslash 2}), \ldots, \text{rank}(\tilde{\mathbf{H}}_{xx\backslash U-1})\right\}$ $- D$. Note this condition is looser than that for applying the ZF precoder, which is one main advantage of the BD precoder [12].

As can be seen in (3.37), $\tilde{\mathbf{V}}_{xx\backslash u}^0$ forms an orthogonal basis for the null space of matrix $\tilde{\mathbf{H}}_{xx\backslash u}$, and its columns therefore form a good candidate for the matrix $\mathbf{M}_{uxx}$. With the BD matrices $\{\mathbf{M}_{uxx}\}_{u=0}^{U_x-1}$, the equivalent channel between the transmitter at x and its u-th associated receiver is given by

$$\hat{\mathbf{H}}_{uxx} = \mathbf{H}_{uxx}\tilde{\mathbf{V}}_{xx\backslash u}^0. \tag{3.38}$$

More importantly, the interference between the receivers that are associated with the same transmitter has already been canceled. This makes the BD based MU-MIMO problem reduce to the SU-MIMO case, where various MIMO techniques were introduced previously in this section.

3.2.3 Distributions of MIMO Signal Power Gains

As shown in Sect. 2.4, it is crucial to understand the distribution of the signal power gain $g_{x_0,d}$ for network performance analysis. In particular, in single-antenna wireless networks with Rayleigh fading, we have $g_{x_0,d} \sim \text{Exp}(1)$, and the derivation of many performance metrics relies on the Laplace transform of the interference and noise. On the other hand, in multi-antenna networks, it is intriguing to know what is the distribution of the signal power gain, and whether tenable analytical approaches can be developed.

Table 3.1 lists some commonly used multi-antenna transmission techniques and channel fading distributions, and the corresponding distributions of $g_{x_0,d}$. It is shown that the gamma distribution is typically encountered in the analysis of multi-antenna networks. In particular, the signal power gain $g_{x_0,d}$ has been proved to be gamma distributed for various MIMO techniques, including MRT [1], partial ZF beamforming and combining [16], MRC [8], SDMA [19], jamming [20], and analog beamforming [18]. For more general multi-antenna transmission strategies, gamma distribution was shown to be an accurate approximation of the channel power gain distribution [21]. A more general yet complicated distribution was derived for g_{x_0} in multi-antenna networks in [22], which incorporates the abovementioned MIMO techniques as

Table 3.1 Typical multi-antenna transmission techniques and corresponding signal power gain distributions.

	Multi-antenna transmission technique ($\mathbf{F}_{0x_0}/\mathbf{W}_{0x_0}$)	Channel fading ($\mathbf{H}_{0x_0x_0}$)	Signal power gain ($g_{x_0,d}$) distribution
Throughput and energy efficiency analysis [14]	MRT	Rayleigh	$\text{Gamma}(N_t, 1)$
Interference coordination [15]	Partial ZF beamforming	Rayleigh	$\text{Gamma}(\max(N_t - N_{x_0}, 1), 1)$
SIMO ad hoc networks [16]	Partial ZF combining	Rayleigh	$\text{Gamma}(N_r - N_{x_0}, 1)$
Spatial multiplexing in ad hoc networks [8]	Maximum ratio combining (MRC)	Rayleigh	$\text{Gamma}(N_r, 1)$
Multi-tier multiuser MIMO HetNets [3]	SDMA	Rayleigh	$\text{Gamma}(N_t - U + 1, 1)$[a]
Physical layer security-aware networks [17]	Jamming and ZF beamforming	Rayleigh	$\text{Gamma}(D, 1)$
Millimeter-wave networks [18]	Analog beamforming	Nakagami	$\text{Gamma}(N_t, 1/N_t)$

[a]The parameters are for each tier in HetNets

The numbers of antennas at the transmitter and receiver sides are denoted as N_t and N_r, respectively, and U denotes the number of served users in SDMA systems. In addition, N_{x_0} represents the number of transmitters that the typical receiver requests to perform interference canceling. Please refer to the corresponding references for more details

special cases, and can characterize more advanced MIMO techniques such as antenna selection. To keep the presentation clean, the analytical framework in this chapter is based on the gamma distribution, denoted as $g_{x_0,d} \sim \text{Gamma}(M, \theta)$, where M and θ are the shape and scale parameters. More importantly, the analytical framework presented in this chapter can be applied to the more general distribution of g_{x_0} introduced in [22], as will be discussed briefly in Sect. 3.3.3.

Next, some basics of gamma distribution are introduced. The pdf of a gamma distributed random variable X is given by

$$f_X(x; M; \theta) = \frac{x^{M-1} e^{-\frac{x}{\theta}}}{\theta^M \Gamma(M)}, \tag{3.39}$$

where M and θ denote the shape and scale parameters, and $\Gamma(M)$ is the gamma function evaluated at M. Therefore, the cdf of the gamma distribution is given by

$$F_X(x; k, \theta) = \int_0^x f(u; M, \theta)\mathrm{d}u = \frac{\gamma\left(M, \frac{x}{\theta}\right)}{\Gamma(M)}, \tag{3.40}$$

where $\gamma(s, x)$ is the lower incomplete gamma function [23, pp. 890]. Accordingly, the mean of X is $\mathbb{E}[X] = M\theta$, while the variance is given by $\text{Var}(X) = M\theta^2$. When M is a positive integer, the cdf has a more explicit expressions as

$$F_X\left(x; k, \theta\right) = 1 - \sum_{i=0}^{M-1} \frac{1}{i!} \left(\frac{x}{\theta}\right)^i e^{-x/\theta}. \tag{3.41}$$

3.3 A General Analytical Framework

In this section, the main challenges of multi-antenna network analysis are first identified. Then a general analytical framework is presented, which applies when the signal power gain is gamma distributed and the interferer's power gains are arbitrarily distributed. As mentioned in Chap. 2, one main dominant theme in network performance analysis is to characterize the distribution of the SINR value, which is the focus of this section.

3.3.1 Challenges in Multi-Antenna Network Analysis

In contrast to single-antenna networks, the deployment of multiple antennas creates the possibility of more advanced communication techniques, such as SDMA, interference alignment, and artificial jamming. To employ these techniques, an accurate characterization of the system performance is required. For example, antenna selection is one way to achieve a satisfactory energy efficiency of MIMO systems. In this way, the relationship between the energy efficiency and the number of active antennas needs to be analytically depicted so that advance optimization techniques can be applied to find the optimal antenna size. However, analytically characterizing the performance of multi-antenna networks is highly difficult. In particular, the power gains for both the desired signal and interference are much more complicated than those in single-antenna ones, due to various adopted MIMO transmission techniques. In the following, we first present the technical challenges in analyzing multi-antenna networks and then compare different approaches for the performance analysis of multi-antenna networks.

There are various performance metrics for wireless networks, e.g., the outage probability, ASE, average throughput, and energy efficiency. As shown in Sect. 2.2, one fundamental task in characterizing these metrics is to calculate the SINR distribution [3, 14, 15, 24], which is equivalent to the coverage probability. Recall that the coverage probability is defined in (2.21) as[3]

$$p_{\mathrm{c}}(\tau) \triangleq \mathbb{P}(\mathrm{SINR} \geq \tau). \tag{3.42}$$

As shown in Sect. 2.4, the coverage probability of single-antenna networks with Rayleigh fading is critically determined by the Laplace transform $\mathscr{L}(s)$ of the aggregated interference. Next, we will show the fundamentals for evaluating the performance of multi-antenna networks, and present several existing approaches to the coverage probability analysis.

According to the SINR expression (3.18),[4] the coverage probability defined in (3.42) can be written as

$$p_{\mathrm{c}}(\tau) = \mathbb{P}\left[g_{x_0} \geq \tau r_0^{\alpha}\left(\sigma_{\mathrm{n}}^2 + I\right)\right], \tag{3.43}$$

where $I = \sum_{x\in\Phi'} g_x \|x\|^{-\alpha}$ is the aggregated interference.

As illustrated in Table 3.1, the gamma distribution is commonly encountered for the signal power gain g_{x_0}. Therefore, it shall be adopted in the main context of this monograph while the extension to more general distributions will be explained in Sect. 3.3.3. According to the ccdf of the gamma distribution, the coverage probability in (3.43) is first rewritten as

$$\begin{aligned} p_{\mathrm{c}}(\tau) &= \mathbb{E}_{r_0}\left\{\sum_{n=0}^{M-1} \frac{(\tau r_0^{\alpha}/\theta)^n}{n!}\mathbb{E}_I\left[(\sigma_{\mathrm{n}}^2 + I)^n e^{-\frac{\tau r_0^{\alpha}}{\theta}(\sigma_{\mathrm{n}}^2+I)}\middle| r_0\right]\right\} \\ &= \mathbb{E}_{r_0}\left[\sum_{n=0}^{M-1} \frac{(-s)^n}{n!}\mathscr{L}^{(n)}(s)\right], \end{aligned} \tag{3.44}$$

[3]For a K-tier HetNet, the coverage probability $p_{\mathrm{c},k}(\tau)$, given that the typical receiver is associated with the k-th tier, can be calculated by (3.42), and the overall coverage probability is then given by $\sum_{k=1}^{K} A_k p_{\mathrm{c},k}(\tau)$, where A_k is the probability that the typical receiver is associated with the k-th tier.

[4]Here we omit the index d for the data stream in order to provide a neat presentation.

where $s \triangleq \tau r_0^\alpha/\theta$, and $\mathscr{L}(s) \triangleq e^{-s\sigma_n^2}\mathbb{E}_I\left[e^{-sI}\middle| r_0\right]$ is the Laplace transform of noise and interference conditioned on the distance r_0.

From (3.44), we observe that, in contrast to single-antenna networks, the coverage probability of multi-antenna networks not only depends on the Laplace transform $\mathscr{L}(s)$ itself but also its higher order derivatives, for which the number of orders is determined by the shape parameter of the gamma distribution. Recall that in single-antenna networks with Rayleigh fading channels, calculating the derivatives of the Laplace transform is not needed, as the signal power gain is exponentially distributed. While calculating the Laplace transform in single-antenna networks entails abundant existing results for various network settings and performance metrics, it is highly nontrivial to derive tractable expressions for the derivatives of the Laplace transform for multi-antenna networks, even for simple network settings. This forms the main challenge in analyzing multi-antenna networks. In the following, we introduce some existing approaches to address this problem.

First-Order Taylor Series Approximation

In [1, 25], by assuming that the Laplace transform has the form

$$\mathscr{L}(s) = \exp\left(-cs^d\right), \tag{3.45}$$

where c and d are parameters determined by different network settings, the n-th derivative of the Laplace transform is calculated by using a first-order Taylor series approximation around $cs^d = 0$. The expression is given by

$$\frac{\mathrm{d}^n}{\mathrm{d}s^n}\mathscr{L}(s) = -c\prod_{m=0}^{n-1}(d-m)s^{d-n}\exp\left(-cs^d\right) + \Theta\left(c^2s^{2d}\right), \tag{3.46}$$

where Θ denotes the asymptotically tight bound. Based on this approximation, tractable analytical results were derived, and key network insights were unraveled. For example, Hunter et al. [1] determined, under small outage constraints, e.g., $\varepsilon < 0.1$, the scaling of the optimal contention density λ_ε (See Sect. 2.4.2) with the number of antennas for different multi-antenna transmission techniques. In particular, we have:

- For ideal sectorized antennas: $\lambda_\varepsilon = \Theta\left(N_\mathrm{t}N_\mathrm{r}\right)$.
- For MRC: $\lambda_\varepsilon = \Theta\left(N_\mathrm{r}^\delta\right)$.
- For MRT and MRC: $\lambda_\varepsilon = \mathscr{O}\left((N_\mathrm{t}N_\mathrm{r})^\delta\right)$, $\lambda_\varepsilon = \Omega\left(\max\{N_\mathrm{t}, N_\mathrm{r}\}^\delta\right)$, where $\mathscr{O}$ and Ω denote the asymptotic upper and lower bounds, respectively.
- for orthogonal space-time block coding (OSTBC): $\lambda_\varepsilon = \Theta\left(N_\mathrm{r}^\delta\right)$.

Upper Bounds

Instead of providing approximations for the derivatives of the Laplace transform, some works tried to avoid directly calculating the high-order derivatives by deriving upper bounds on the outage probability. In [16], the outage probability is upper bounded by

$$p_o(\tau) = \mathbb{P}\left(\frac{1}{\mathrm{SINR}} > \frac{1}{\tau}\right) \leq \tau \mathbb{E}\left[\frac{1}{\mathrm{SINR}}\right] \tag{3.47}$$

$$= \tau \mathbb{E}\left[r_0^\alpha I + r_0^\alpha \sigma_n^2\right] \mathbb{E}\left[\frac{1}{g_{x_0}}\right]$$

$$= \frac{\tau}{\theta(M-1)} \mathbb{E}\left[r_0^\alpha I + r_0^\alpha \sigma_n^2\right], \tag{3.48}$$

where (3.47) is due to Markov's inequality, and (3.48) is because $1/g_{x_0}$ is an inverse-gamma distributed random variable. In this upper bound, what we need to calculate is only the expectation of the aggregated interference, which can be readily computed according to Campbell's theorem for sums, i.e., Theorem 2.2, and therefore, it is much simplified compared with computing the derivatives of the Laplace transform.

In [26], the calculation of the derivatives of the Laplace transform is avoided by directly introducing a tight upper bound for the cdf of a gamma random variable g_{x_0} with parameter the $\theta = 1/M$, given by

$$\mathbb{P}\left(g_{x_0} < x\right) < \left(1 - e^{-ax}\right)^M, \tag{3.49}$$

where $a = M(M!)^{-\frac{1}{M}}$. By applying this upper bound, the coverage probability is hence lower bounded by

$$p_c(\tau) > \mathbb{E}_{r_0}\left[1 - \mathbb{E}_I\left\{\left[1 - e^{-a\tau r_0^\alpha(\sigma_n^2 + I)}\right]^M\right\}\right]$$

$$= \mathbb{E}_{r_0}\left[\sum_{m=1}^{M}(-1)^{m+1}\binom{M}{m}\mathbb{E}_I\left[e^{-ma\tau r_0^\alpha(\sigma_n^2+I)}\right]\right] \tag{3.50}$$

$$= \mathbb{E}_{r_0}\left[\sum_{m=1}^{M}(-1)^{m+1}\binom{M}{m}e^{-ma\tau r_0^\alpha \sigma_n^2}\mathscr{L}_I(s_m)\right],$$

where $s_m \triangleq ma\tau r_0^\alpha$. Now, what are required for evaluating the coverage probability are the Laplace transforms with M different arguments, with is much easier than deriving M higher order derivatives. This technique has been widely adopted for analyzing mm-wave networks with Nakagami-M fading, e.g., in [26, 27].

Exact Analytical Results

There are also some works that tried to derive exact analytical results for the n-th derivatives of the Laplace transform and the coverage probability. In [28], using Faà di Bruno's lemma [29] and assuming that the Laplace transform has the form

$$\mathscr{L}(s) = e^{\eta(s)}, \tag{3.51}$$

where $\eta(s)$ is the exponent of the Laplace transform, the n-th derivative of the Laplace transform is calculated. The expression is given by

$$\frac{\mathrm{d}^n}{\mathrm{d}s^n}\mathscr{L}(s) = \mathscr{L}(s) \sum_{\bar{m}\in\mathscr{M}(n)} \frac{n!}{\prod_i (m_i!(i!)^{m_i})} \prod_{i=1}^{n} \left(\eta^{(i)}(s)\right)^{m_i}, \tag{3.52}$$

where

$$\mathscr{M}(n) = \left\{\bar{m} = (m_1, m_2, \ldots, m_n) : \sum_{i=1}^{n} i m_i = n\right\}. \tag{3.53}$$

To numerically evaluate the expression in (3.52), one should first solve a Diophantine equation in (3.53), which is computationally heavy when n is large.

While expressed in a complicated form and difficult to evaluate, the analysis in [28] did lead to some interesting results. For example, by expressing the coverage probability of multi-antenna HetNets in a closed form, it was revealed that the coverage probability is no longer scale invariant in multi-antenna HetNets even for the interference-limited case. Assuming arbitrary selection bias for each tier, simple expressions for downlink coverage and rate were also derived. Moreover, for coverage maximization, the required selection bias for each tier is found and given in a closed form.

In [8, 22], the coverage probability of ad hoc networks was derived. The desired signal power gain g_{x_0} was proved to be gamma distributed for typical MIMO techniques over Rayleigh fading channels, e.g., MRC, ZF, and OSTBC. By leveraging the result in [30, Eq. (113)] and the Dobiński's formula [31], the coverage probability is expressed as

$$\begin{aligned} p_{\mathrm{c}}(\tau) =& \frac{(-1)^{M-1}\mathscr{L}(s)e^{-\tau/\theta}}{\Gamma(M)} \sum_{l=0}^{M-1} \binom{M-1}{l} \left(-\frac{\tau}{\theta}\right)^l \\ & \times \sum_{i=0}^{M-l-1} S(M-l, i+1)\delta^i \sum_{j=0}^{i} s(i,j) \left[\log \mathscr{L}(s)\right]^j, \end{aligned} \tag{3.54}$$

where

$$\mathscr{L}(s) = \pi\lambda\Gamma(1-\delta) r_0^2 \left(\frac{\tau}{\theta}\right)^{\delta} \mathbb{E}_g\left[g^{\delta}\right], \tag{3.55}$$

and g is distributed as g_x for all $x \in \Phi$. In addition, $S(n,k)$ and $s(n,k)$ denote the Stirling numbers of the first and second kind, respectively. Following this result, the transmission capacity of ad hoc networks with three commonly adopted MIMO techniques was compared in [8]. It was shown that multi-antenna schemes with a simple decentralized slotted ALOHA medium access control outperform even idealized single-antenna networks in various practical scenarios. This work was extended to more general linear MIMO transmission schemes in [22]. However, the bulky form of the result in (3.54) is the main obstacle to yield insights for network design and optimization. For example, the nested sums prevent us from investigating the impact of the shape parameter M (which is closely related to the number of antennas as later we shall see), as both the numbers of terms in the nested sums and the terms themselves with the complicated Stirling numbers are determined by M.

As a brief summary, the abovementioned approaches to evaluating multi-antenna networks, either approximately or with very bulky expressions, are only suitable for numerical evaluation or qualitative compassion, but not able to yield insights for network design and optimization quantitatively. Therefore, it is crucial to develop a unified framework to analyze multi-antenna networks based on which key system insights can be revealed.

3.3.2 Analytical Framework

In this subsection, we introduce an analytical framework for the tractable analysis of multi-antenna networks. As we mentioned above, the signal power gain is assumed to be gamma distributed as $g_{x_0} \sim \text{Gamma}(M, \theta)$. In the general framework, we assume that the interferer's power gain g_x is with an arbitrary distribution. It is not necessary that all $\{g_x\}_{x \in \Phi'}$ are distributed according to the same distribution. Instead, they can be several families of random variables over different point processes, as shown in the following example.

Example 3.4 (Network with multiple types of interferers) We assume that $\left\{(g_x)_{x \in \Phi'_j}\right\}_{j=1}^J$ are J families of nonnegative random variables that are independent and identically distributed according to arbitrary distributions, for which the $\frac{2}{\alpha}$-th moments exist. This model not only reflects the general multi-tier HetNet setting but also applies when the interferers have different channel power gain distributions in a single-tier network. According to the PGFL of PPP in (2.38), the conditional Laplace transform $\mathscr{L}(s)$ can be expressed in a general exponential form as

$$\mathscr{L}(s) = \exp\left\{ - s\sigma_{\text{n}}^2 - 2\pi \sum_{j=1}^{J} \lambda_j \int_{l_j(r_0)}^{\infty} \left(1 - \mathbb{E}_{g_j}[\exp(-sg_j v^{-\alpha})]\right) v\mathrm{d}v \right\} \tag{3.56}$$

where λ_j is the density[5] of Φ'_j, the interferers' power gain is denoted as g_j that is identically distributed as all the $(g_x)_{x \in \Phi'_j}$.

As the exponent of the Laplace transform $\mathscr{L}(s)$ will commonly appear in the following discussion, similar to [28], we express the Laplace transform in the form

$$\mathscr{L}(s) = \exp\{\eta(s)\}. \tag{3.57}$$

[5] For network models incorporating load awareness [14, 32], the activation of transmitters can be reflected in the density λ_j.

We use $\eta(s)$ to denote the exponent of the Laplace transform, which is called the *log-Laplace transform*. Note that, according to (2.42), the Laplace transform of a sum over a homogeneous PPP can be generally expressed as an exponential function. Moreover, for other more complicated point processes, e.g., inhomogeneous PPP and Poisson hole process introduced in Sect. 2.1.3, the Laplace transforms are also exponential functions. Therefore, the form in (3.57) is general and incorporates a bunch of practical network models based on various point processes.

While there exist some approaches to calculate the n-th derivative of a general exponential function for the Laplace transform $\mathscr{L}(s)$, e.g., via Faà di Bruno's formula [29] or Bell polynomials [23], a direct computation of the derivatives leads to unwieldy expressions [8, 22, 28, 33], which cannot be efficiently evaluated and fail to reveal key system insights. In contrast, as we will see, the presented framework enables tractable results for the coverage probability of multi-antenna networks, and, more importantly, helps reveal key network insights. There are three ingredients in this framework that lead to tractable results, which will be introduced sequentially in this section. Here we introduce the first ingredient, i.e., **instead of working with the Laplace transform $\mathscr{L}(s)$ directly, we analyze the log-Laplace transform $\eta(s)$.**

Remark 3.1 While we have given a specific example of the Laplace transform in Example 3.4, the framework to be presented does not depend on the particular form of the log-Laplace transform, and it can be readily extended to other network models, for example, where the transmitters are spatially distributed according to other point processes [34], or the multi-slope path loss model [35]. One can first determine the log-Laplace transform $\eta(s)$ according to the network model, and then the analytical framework can be applied. In addition, as established in [36], the SIR coverage probability of cellular non-Poisson models is well approximated by $p_{\mathrm{c}}(\tau/G)$, where $p_{\mathrm{c}}(\tau)$ is the coverage probability of the cellular Poisson model and G is a gain factor that depends on the geometry of the non-Poisson model. Hence, the results in this chapter permit a simple approximation of the coverage probabilities for any stationary and ergodic point process model.

First, the recursive relation between the derivatives of the Laplace transform is revealed in the following lemma.

Lemma 3.1 *Defining* $p_n = \frac{(-s)^n}{n!}\mathscr{L}^{(n)}(s)$*, there exist recursive relations between* $\{p_n\}_{n=0}^{\infty}$*, given by*

$$p_n = \sum_{i=0}^{n-1} \frac{n-i}{n} t_{n-i} p_i, \tag{3.58}$$

where

$$t_k = \frac{(-s)^k}{k!} \eta^{(k)}(s). \tag{3.59}$$

Proof First, it is obvious that $p_0 = \mathscr{L}(s) = e^{\eta(s)}$ and $\mathscr{L}^{(1)}(s) = \eta^{(1)}(s)\mathscr{L}(s)$. According to the formula of Leibniz for the n-th derivative of the product of two functions [29], we have

$$\mathscr{L}^{(n)}(s) = \frac{\mathrm{d}^{n-1}}{\mathrm{d}s}\mathscr{L}^{(1)}(s) = \sum_{i=0}^{n-1}\binom{n-1}{i}\eta^{(n-i)}(s)\mathscr{L}^{(i)}(s), \tag{3.60}$$

followed by

$$\frac{(-s)^n}{n!}\mathscr{L}^{(n)}(s) = \sum_{i=0}^{n-1}\frac{n-i}{n}\frac{(-s)^{(n-i)}}{(n-i)!}\eta^{(n-i)}(s)\frac{(-s)^i}{i!}\mathscr{L}^{(i)}(s), \tag{3.61}$$

which completes the proof by applying the definition that $p_n = \frac{(-s)^n}{n!}\mathscr{L}^{(n)}(s)$.

According to the recursive relations in (3.58) and the fact that $p_0 = \mathscr{L}(s)$, the only factors we need to calculate to obtain $\{p_n\}_{n=1}^{M-1}$ are the coefficients $\{t_k\}_{k=0}^{M-1}$, which are related to the derivatives of $\eta(s)$. So the main task is shifted from calculating the derivatives of $\mathscr{L}(s)$ to deriving those of $\eta(s)$. As shown in extensive existing works [1, 3, 8, 14–17, 20, 22, 25, 28, 37–39], obtaining a closed-form expression for $\eta^{(n)}(s)$ is generally much easier than for $\mathscr{L}^{(n)}(s)$, which will be further demonstrated in this chapter. Following Lemma 3.1, a finite sum representation of the coverage probability is given in the following theorem.

Theorem 3.1 (Finite sum representation of the coverage probability) *The coverage probability* (3.43) *is given by*

$$p_\mathrm{c}(\tau) = \mathbb{E}_{r_0}\left[\sum_{n=0}^{M-1} p_n\right]. \tag{3.62}$$

where $\{p_n\}_{n=0}^{M-1}$ *are given in Lemma 3.1.*

Proof The result follows from (3.44) and the definition of p_n in Lemma 3.1.

Remark 3.2 The main merit of this representation is that it leads to valuable system insights. For example, the impact of the shape parameter M in the gamma distribution, which is typically related to the antenna size, is clearly illustrated by this finite sum representation. We define $\bar{p}_n \triangleq \mathbb{E}_{r_0}[p_n]$ to simplify the presentation. In particular, as shown in Table 3.1 and the references therein, the interferers' power gains $g_{x,j}$ are typically independent of M for various MIMO transmission techniques, and so are $\{\bar{p}_n\}_{n=0}^{M-1}$. When M increases, e.g., from M to $M + \Delta$, the number of terms in the sum increases, and the variation of the coverage probability is directly related to the coefficients $\{\bar{p}_n\}_{n=M}^{M+\Delta-1}$. This property will be leveraged to reveal the impact of the antenna size in Sect. 3.5.

From both (3.44) and (3.62), it is apparent that the main challenge in evaluating the coverage probability is to derive a tractable expression for $\{p_n\}_{n=0}^{M-1}$. With Theorem 3.1, we need to calculate $\{p_n\}_{n=0}^{M-1}$ in a recursive manner, which is still tedious. Next, we derive more explicit expressions for $\{p_n\}_{n=0}^{M-1}$, assuming that we have obtained $\{t_k\}_{k=0}^{M-1}$.

To this end, we introduce the second ingredient in the analytical framework that results in tractable results. In particular, **two power series are introduced to help the derivation**, which are defined as follows:

$$T(z) \triangleq \sum_{n=0}^{\infty} t_n z^n, \quad P(z) \triangleq \sum_{n=0}^{\infty} p_n z^n. \tag{3.63}$$

The coefficients of the power series are given in (3.59) and (3.58), respectively. While only the first M coefficients are required for obtaining the coverage probability, it is shown in the following lemma that introducing the power series helps reveal a more explicit (non-recursive) relation between $\{p_n\}_{n=0}^{M-1}$ and $\{t_k\}_{k=0}^{M-1}$ than (3.58).

Lemma 3.2 *The power series $P(z)$ is related to $T(z)$ as*

$$P(z) = e^{T(z)}. \tag{3.64}$$

Proof It is straightforward to show that $T^{(1)}(z) = \sum_{n=0}^{\infty}(n+1)t_{n+1}z^n$ and $P^{(1)}(z) = \sum_{n=0}^{\infty} np_n z^{n-1}$. We then have the following equality:

$$T^{(1)}(z)P(z) = \sum_{n=0}^{\infty}\sum_{i=0}^{n-1}(n-i)t_{n-i}p_i z^{n-1}. \tag{3.65}$$

Combined with (3.58), we obtain the differential equation

$$P^{(1)}(z) = T^{(1)}(z)P(z), \tag{3.66}$$

whose solution is given by (3.64).

Recall that the coverage probability is determined by the first M coefficients of the power series $P(z)$. With Lemma 3.2, these coefficients can be represented by the coefficients of $T(z)$ in closed forms, which is, however, too complicated for further investigations. To address this problem, we introduce the third ingredient in the analytical framework. In particular, **we present an important relation between power series and a special form of matrix, i.e., the matrix representation of power series** [40, Ch. 1]. For a power series

$$X(z) = \sum_{n=0}^{\infty} x_n z^n, \tag{3.67}$$

we define an infinite lower triangular Toeplitz matrix with its coefficients, given by

$$\mathbf{X}_{\infty} = \begin{bmatrix} x_0 & & & & \\ x_1 & x_0 & & & \\ x_2 & x_1 & x_0 & & \\ x_3 & x_2 & x_1 & x_0 & \\ \vdots & \vdots & \vdots & & \ddots \end{bmatrix}, \tag{3.68}$$

and we describe the correspondence between the power series $X(z)$ and the infinite lower triangular Toeplitz matrix $\mathbf{X}_{\infty}$ as

$$X(z) \mapsto \mathbf{X}_{\infty}. \tag{3.69}$$

In the area of abstract algebra, an *isomorphism* is a correspondence relation (mapping) between objects expressing the equality of their structures in some sense. For example, let $\log : (\mathbb{R}^+, \times) \mapsto (\mathbb{R}, +)$ be the mapping: $x \mapsto \log x$, where $(\mathbb{R}^+, \times)$ and $(\mathbb{R}, +)$ are the multiplicative group of positive real numbers and additive group of real numbers, respectively. Then, according to the logarithm law, i.e., $\log(xy) = \log x + \log y$, the logarithm log is a group isomorphism. The matrix representation of the introduced power series will take effect via the following important result.

Lemma 3.3 ([40, Ch. 1.3]) *The mapping* $\mapsto$ *from the set of power series onto the set of infinite lower triangular Toeplitz matrices is an isomorphism.*

For instance,

$$\text{if } X(z) \mapsto \mathbf{X}_{\infty}, \quad Y(z) \mapsto \mathbf{Y}_{\infty}, \qquad \text{then } \begin{aligned} X(z) + Y(z) &\mapsto \mathbf{X}_{\infty} + \mathbf{Y}_{\infty} \\ X(z)Y(z) &\mapsto \mathbf{X}_{\infty}\mathbf{Y}_{\infty}. \end{aligned} \tag{3.70}$$

It is easy to check that the product of two such matrices always exists and is still a lower triangular Toeplitz matrix. If we define the n-th square submatrix of $\mathbf{X}_{\infty}$ as $\mathbf{X}_n$, given by

$$\mathbf{X}_n = \begin{bmatrix} x_0 & & & & \\ x_1 & x_0 & & & \\ x_2 & x_1 & x_0 & & \\ \vdots & & & \ddots & \\ x_{n-1} & \cdots & x_2 & x_1 & x_0 \end{bmatrix}, \tag{3.71}$$

then the product of two submatrices is equal to the n-th square submatrix of the product, which is mathematically given by

$$\mathbf{X}_n\mathbf{Y}_n = (\mathbf{X}_{\infty}\mathbf{Y}_{\infty})_n . \tag{3.72}$$

In other words, the property of the lower triangular Toeplitz matrix guarantees that the first n entries in the first columns of the product $\mathbf{X}_\infty \mathbf{Y}_\infty$ only depend on the first n elements in the first column of $\mathbf{X}_\infty$ and $\mathbf{Y}_\infty$. Correspondingly, according to the isomorphism in Lemma 3.3, the first n coefficients of the series product $X(z)Y(z)$ are only related to the first n coefficients of $X(z)$ and $Y(z)$. In addition, this property can also be easily proved for the sum of power series.

Next, we shall introduce an important notation in this chapter, i.e., the induced ℓ_1-norm of a matrix $\mathbf{X} \in \mathbb{C}^{m\times n}$. It is defined as

$$\|\mathbf{X}\|_1 = \max_{1\leq j\leq n} \sum_{i=1}^{m} |x_{ij}|, \tag{3.73}$$

which is simply the maximum absolute column sum of the matrix.

Based on Lemma 3.2, Lemma 3.3, and the definition in (3.73), an explicit expression for the coverage probability is given in Theorem 3.1, which is more tractable than the result in Theorem 3.1.

Theorem 3.2 (ℓ_1-Toeplitz matrix representation of the coverage probability) *The coverage probability* (3.43) *is given by*

$$p_c(\tau) = \mathbb{E}_{r_0}\left[\left\|e^{\mathbf{T}_M}\right\|_1\right], \tag{3.74}$$

where $\mathbf{T}_M$ is the following $M \times M$ lower triangular Toeplitz matrix:

$$\mathbf{T}_M = \begin{bmatrix} t_0 & & & & \\ t_1 & t_0 & & & \\ t_2 & t_1 & t_0 & & \\ \vdots & & & \ddots & \\ t_{M-1} & \cdots & t_2 & t_1 & t_0 \end{bmatrix}, \tag{3.75}$$

and its nonzero entries are determined by (3.59).

Proof We first present the correspondent infinite lower triangular Toeplitz matrix $\mathbf{P}_\infty$ of the power series $P(z)$. Since $P(z) = e^{T(z)}$, according to Lemma 3.3,

$$P(z) = e^{T(z)} = \sum_{n=0}^{\infty} \frac{[T(z)]^n}{n!} \mapsto \mathbf{P}_\infty = \sum_{n=0}^{\infty} \frac{\mathbf{T}_\infty^n}{n!} = e^{\mathbf{T}_\infty}. \tag{3.76}$$

The sum $\sum_{p=0}^{M-1} p_n$ in (3.62) is the first column sum of $(\mathbf{P}_\infty)_M$, which can be rewritten as

Methodology 1 Main Steps to Apply the Analytical Framework

1: Derive the conditional Laplace transform $\mathscr{L}(s)$ according to 3.56 for the given distributions of $\{g_j\}_{j=1}^{J}$ and the specific point processes for the interfering transmitters $\{\Phi'_j\}_{j=1}^{J}$;
2: Calculate the n-th $(1 \le n \le M-1)$ derivatives of the log-Laplace transform $\eta(s)$ to populate the entries $\{t_n\}_{n=0}^{M-1}$ in the matrix $\mathbf{T}_M$ according to 3.59;
3: Express the coverage probability $p_c(\tau)$ with Theorem 3.2.

$$
\begin{aligned}
(\mathbf{P}_\infty)_M &= \left(\sum_{n=0}^{\infty} \frac{\mathbf{T}_\infty^n}{n!}\right)_M = \sum_{n=0}^{\infty} \frac{\left(\mathbf{T}_\infty^n\right)_M}{n!} \\
&= \sum_{n=0}^{\infty} \frac{\mathbf{T}_M^n}{n!} = e^{\mathbf{T}_M},
\end{aligned}
\tag{3.77}
$$

where (3.77) applies the property in (3.72). It can be readily shown from Lemma 3.1 that $p_n > 0$ for $0 \le n \le M-1$. Therefore, by the definition of the induced ℓ_1 matrix norm, the sum $\sum_{p=0}^{M-1} p_n$ in (3.62) can be rewritten as the induced ℓ_1-norm of the submatrix $e^{\mathbf{T}_M}$, which completes the proof.

Table 3.2 lists the commonly adopted analytical approaches for multi-antenna networks. Compared with the approximations in [1, 16, 25, 26, 37] and complicated expressions in [8, 20, 22, 28, 38, 39], the ℓ_1-Toeplitz matrix representation in (3.74) provides a more compact form for the coverage probability. More importantly, it enables us to leverage various powerful tools from linear algebra, especially some nice properties of lower triangular Toeplitz matrices, to provide insightful design guidelines for network optimization. Such properties will be demonstrated later in Chaps. 4 and 5. Intuitively, the matrix representation is also a more natural way to express results for multi-antenna systems with vector channels, although the matrix form presented here is different from those for link-level analysis of MIMO systems.

When applying Theorem 3.2 to specific multi-antenna networks, the only parameters to be determined are the nonzero entries $\{t_n\}_{n=0}^{M-1}$ in the matrix $\mathbf{T}_M$, and the main steps for applying the framework are summarized as Methodology 1. It is a three-step methodology for analyzing multi-antenna networks, and all the analysis in the remaining parts of this monograph shall follow this rule of thumb to provide tractable results and reveal key network insights.

Table 3.2 Comparison of different analytical approaches for multi-antenna networks.

Approach	Tractability	Accuracy
First-order Taylor series approximation [1, 25]	✓✓	✓✓
Upper bounds [16, 26]	✓✓✓	✓
Exact analytical results [8, 22, 28]	✓	✓✓✓
ℓ_1-Toeplitz matrix representation	✓✓✓	✓✓✓

Table 3.3 The use of Theorems 3.1 and 3.1 in Sects. 3.4 and 3.5.

	Corollaries 3.1 and 3.2 Propositions 3.1, 3.2, and 3.3	Corollary 3.4 Propositions 3.4, 3.5, and 3.6	Corollary 3.3
Theorem 3.1		✓	✓
Theorem 3.2	✓		✓

Remark 3.3 Although an additional expectation over r_0 is needed in (3.74) when the distance between the typical receiver and its associated transmitter is a random variable, e.g., in cellular networks, in the next section we will show that closed-form expressions are available via the framework. As listed in Table 3.3, in the remainder of this chapter, Theorems 3.1 and 3.2 will be utilized to analyze multi-antenna networks in specific settings.

3.3.3 More General Distributions for the Signal Power Gain

While we mainly consider gamma distributed signal power gains, the analytical framework applies to more general cases, as illustrated in this subsection. We consider a more general form of the pdf of g_{x_0} that may be encountered in multi-antenna wireless networks, which is given by [22, Eq. (10)]

$$f_{g_{x_0}}(u) = \sum_{i \in \mathscr{I}} e^{-\phi_i u} \sum_{j \in \mathscr{J}} \varphi_{i,j} u^j, \tag{3.78}$$

where $\mathscr{I}$, $\mathscr{J} \subset \mathbb{N}_0$ and $\phi_i, \varphi_{i,j} \in \mathbb{R}$ are model parameters. In addition, various special cases of this pdf with different MIMO transmission techniques, e.g., transmit antenna selection with ZF receivers and open-loop spatial multiplexing with ZF receivers, are specified in [22, Table I]. According to (3.78), the ccdf of g_{x_0} is given by

$$F^c_{g_{x_0}}(u) = 1 - \sum_{i \in \mathscr{I}} \sum_{j \in \mathscr{J}} \frac{\varphi_{i,j} j!}{\phi_i^{j+1}} + \sum_{i \in \mathscr{I}} \sum_{j \in \mathscr{J}} \frac{\varphi_{i,j} j!}{\phi_i^{j+1}} \sum_{k=0}^{j} e^{-\phi_i u} \frac{(\phi_i u)^k}{k!}. \tag{3.79}$$

The framework is also applicable to this general form of pdf, as shown in the following theorem.

Theorem 3.3 (Generalized ℓ_1-Toeplitz matrix representation of the coverage probability) *With the generalized distribution* (3.78) *of the desired signal power gain* g_{x_0}*, the coverage probability is given by*

$$p_c(\tau) = 1 - \sum_{i\in\mathcal{I}}\sum_{j\in\mathcal{J}} \frac{\varphi_{i,j} j!}{\phi_i^{j+1}} + \sum_{i\in\mathcal{I}}\sum_{j\in\mathcal{J}} \frac{\varphi_{i,j} j!}{\phi_i^{j+1}} \mathbb{E}_{r_0}\left[\left\| e^{\mathbf{T}_{i,j+1}} \right\|_1\right], \tag{3.80}$$

where $\mathbf{T}_{i,j+1}$ *in the* i*-th term in the sum denotes a* $(j+1)\times(j+1)$ *lower triangular Toeplitz matrix as*

$$\mathbf{T}_{i,j+1} = \begin{bmatrix} t_{i,0} & & & & \\ t_{i,1} & t_{i,0} & & & \\ t_{i,2} & t_{i,1} & t_{i,0} & & \\ \vdots & & & \ddots & \\ t_{i,j} & \cdots & t_{i,2} & t_{i,1} & t_{i,0} \end{bmatrix}. \tag{3.81}$$

Furthermore, the nonzero entries in $\mathbf{T}_{i,j+1}$ *are given by*

$$t_{i,k} = \frac{(-s_i)^k}{k!}\eta^{(k)}(s_i), \quad 0 \le k \le j, \tag{3.82}$$

where $s_i = \tau r_0^{\alpha}\phi_i$.

Proof The proof follows a similar procedure to that of Theorem 3.2.

In addition, Theorem 3.1 can also be generalized as follows to help reveal key network insights.

Theorem 3.4 (Generalized finite sum representation of the coverage probability) *With the generalized distribution* (3.78) *of the desired signal power gain* g_{x_0}*, the coverage probability is given by*

$$p_c(\tau) = 1 - \sum_{i\in\mathcal{I}}\sum_{j\in\mathcal{J}} \frac{\varphi_{i,j} j!}{\phi_i^{j+1}} + \sum_{i\in\mathcal{I}}\sum_{j\in\mathcal{J}} \frac{\varphi_{i,j} j!}{\phi_i^{j+1}} \mathbb{E}_{r_0}\left[\sum_{n=0}^{j} p_{i,n}\right], \tag{3.83}$$

where $\{p_{i,n}\}_{n=0}^{M-1}$ *are given by*

$$p_{i,n} = \sum_{k=0}^{n-1} \frac{n-k}{n} t_{i,n-k} p_{i,k}, \tag{3.84}$$

with $\{t_{i,k}\}_{k=0}^{j}$ *given in* (3.82).

Proof The proof follows a similar procedure to that of Theorem 3.1.

Note that the results in Theorems 3.2 and 3.1 corresponding to the gamma distribution Gamma(M,θ) are special cases of (3.80) and (3.83), respectively, with

$\mathscr{I} = \{0\}$, $\mathscr{J} = \{M-1\}$, $\phi_0 = \frac{1}{\theta}$, and $\varphi_{0,M-1} = \frac{1}{\theta^M \Gamma(M)}$. In this chapter, to keep the presentation concise and easy to follow, we use the gamma distribution to present the main context, but all the results are equally applicable to the general pdf in (3.78).

3.3.4 Single-Antenna Versus Multi-Antenna Networks

Here we show that the presented analytical framework incorporates the single-antenna network as a special case. Assuming Rayleigh fading, the signal power gain is exponentially distributed in the single-antenna case, i.e., $M = \theta = 1$. In this way, expression (3.74) in Theorem 3.1 (or (3.62) in Theorem 1) simplifies to

$$p_c(\tau) = \int_0^\infty f_{r_0}(r)\mathscr{L}(s)\mathrm{d}r, \tag{3.85}$$

which is exactly the classic result (2.55) for single-antenna networks introduced in Sect. 2.4. Note that, for single-antenna networks, the main task to derive the coverage probability is to derive the conditional Laplace transform $\mathscr{L}(s)$. It has been shown in [41] that, under various assumptions for the interferers' power gain, $\mathscr{L}(s)$ (equivalently $\eta(s)$) can be derived in closed forms. This in turn makes it possible to express the coverage probability in a closed form.

When it comes to multi-antenna networks, Theorem 3.1 is compatible with any forms of $\eta(s)$. Furthermore, with the gamma distributed signal power gain, according to Methodology 1, the only additional task compared with single-antenna networks is to calculate $M-1$ derivatives of $\eta(s)$, which does not introduce much computational complexity and thus preserves the tractability. This means that many manipulation tricks and steps developed for single-antenna networks, e.g., derivation techniques listed in [42, Sect. III], can be adopted to the multi-antenna case. The tractability and effectiveness of the framework will be illustrated in the next section by developing new analytical results for general ad hoc and cellular networks.

3.4 Coverage Analysis of General Multi-Antenna Networks

In this section, to illustrate the effectiveness of the analytical framework, we apply it to general cellular and ad hoc networks. By leveraging the ℓ_1-Toeplitz matrix representation in Theorem 3.2, tractable expressions for the coverage probability are provided. Single-tier networks are considered in this part to keep the presentation neat, but the derivation can be extended to general HetNets by calculating the Laplace transform according to (3.56), and the details can be found in Chap. 5. Furthermore, since wireless networks are interference-limited, we focus on the SIR distribution instead of SINR.

3.4.1 Cellular Networks

In the cellular network model, the pdf of the distance r_0 between the typical user and the serving BS is given by (2.60), and the SIR is expressed as

$$\mathrm{SIR} = \frac{g_{x_0} r_0^{-\alpha}}{\sum_{x\in\Phi\backslash\{x_0\}} g_x \|x\|^{-\alpha}}. \tag{3.86}$$

Since the nearest BS is part of the PPP Φ consisting of all the transmitters, the set of interfering BSs $\Phi' = \Phi\backslash\{x_0\}$ forms a PPP on $\mathbb{R}^2\backslash b(0, r_0)$ conditioned on $x_0 \in \Phi$. Recall that we assumed that g_{x_0} is a gamma distributed random variable, i.e., $g_{x_0} \sim \mathrm{Gamma}(M, \theta)$. Let g be a random variable identically distributed as all the $(g_x)_{x\in\Phi'}$, which shall be frequently used in this chapter. The coverage probability for this model is given in the following result, with a detailed proof.

Proposition 3.1 *When the locations of BSs are modeled as a PPP, and the nearest-BS association is adopted in the cellular network, the SIR coverage probability is given by*

$$p_c(\tau) = \left\|\mathbf{C}_M^{-1}\right\|_1, \tag{3.87}$$

with the nonzero entries in the lower triangular Toeplitz matrix $\mathbf{C}_M$ *as*

$$c_n = \frac{\delta}{\delta - n}\frac{(\tau/\theta)^n}{n!}\mathbb{E}_g\left[g^n {}_1F_1\left(n-\delta; n+1-\delta; -\frac{\tau}{\theta}g\right)\right], \tag{3.88}$$

for $0 \leq n \leq M-1$.

Proof A detailed proof is provided here to illustrate the main steps in applying the framework for the coverage analysis. The proofs for the remaining results follow similar steps and are therefore diverted to the appendix.

We first simplify the expression in Theorem 3.2 under the cellular network model. According to the three-step approach of applying Theorem 3.2 as presented in Methodology 1, we first calculate the log-Laplace transform as

$$\begin{aligned}\eta(s) &= -2\pi\lambda\int_{r_0}^{\infty}\left(1-\mathbb{E}_g[\exp(-sgv^{-\alpha})]\right)v\mathrm{d}v \\ &= \pi\lambda r_0^2 + \pi\lambda\delta s^{\delta}\mathbb{E}_g\left[g^{\delta}\gamma(-\delta, sr_0^{-\alpha}g)\right] \qquad (3.89)\\ &= \pi\lambda r_0^2 - \pi\lambda r_0^2\mathbb{E}_g\left[{}_1F_1\left(-\delta; 1-\delta; -sr_0^{-\alpha}g\right)\right], \qquad (3.90)\end{aligned}$$

where (3.89) can be derived from [2, Eq. (4)] by changing variables $v^{-\alpha} \to y$, and step (3.90) applies the identity $\gamma(s,x) \equiv \frac{x^s}{s}{}_1F_1(s, s+1, -x)$ [23, Sect. 6.45]. Then, the second step is to calculate the n-th derivatives of $\eta(s)$. By utilizing the derivatives

$$\frac{\mathrm{d}^n}{\mathrm{d}z^n}{}_1F_1\left(a; b; z\right) = \frac{\prod_{i=0}^{n-1}(a+i)}{\prod_{i=0}^{n-1}(b+i)}{}_1F_1\left(a+n; b+n; z\right), \tag{3.91}$$

the nonzero entries in $\mathbf{T}_M$ in (3.75) are determined by (3.59), i.e.,

$$\begin{aligned} t_n &= \frac{(-s)^n}{n!}\eta^{(n)}(s) \\ &= -\pi\lambda r_0^2 \frac{\delta}{\delta - n}\frac{(\tau/\theta)^n}{n!} \times \left\{\mathbb{E}_g\left[g^n {}_1F_1\left(n-\delta; n+1-\delta; -\frac{\tau}{\theta}g\right)\right] - \mathbb{1}(n=0)\right\} \\ &= -\pi\lambda r_0^2 \left[c_n - \mathbb{1}(n=0)\right], \end{aligned} \tag{3.92}$$

where $\{c_n\}_{n=0}^{M-1}$ are given in (3.88) and $\mathbb{1}(\cdot)$ denotes the indicator function. The coverage probability is evaluated following (3.74) as

$$p_c(\tau) = \int_0^\infty 2\pi\lambda r e^{-\pi\lambda r^2} \left\| e^{\mathbf{T}_M} \right\|_1 \mathrm{d}r. \tag{3.93}$$

This formula can be further simplified into a closed form by defining a power series similar to (3.63), i.e., $C(z) = \sum_{n=0}^{\infty} c_n z^n$. According to (3.92), we have

$$\begin{aligned} T(z) = \sum_{n=0}^{\infty} t_n z^n &= \pi\lambda r_0^2 \left(1 - c_0 - \sum_{n=1}^{\infty} c_n z^n\right) \\ &= \pi\lambda r_0^2 \left[1 - C(z)\right]. \end{aligned} \tag{3.94}$$

To help the derivation, another power series $\bar{P}(z) = \sum_{n=0}^{\infty} \bar{p}_n z^n$ is defined as

$$\bar{P}(z) \triangleq \mathbb{E}_{r_0}\left[P(z)\right]. \tag{3.95}$$

Hence, the power series $\bar{P}(z)$ is written as

$$\begin{aligned} \bar{P}(z) &= \mathbb{E}_{r_0}\left[P(z)\right] \\ &= \mathbb{E}_{r_0}\left[e^{T(z)}\right] = \int_0^\infty 2\pi\lambda r e^{-\pi\lambda r^2} e^{T(z)} \mathrm{d}r \qquad (3.96) \\ &= \int_0^\infty 2\pi\lambda r e^{-\pi\lambda C(z) r^2} \mathrm{d}r = \frac{1}{C(z)}, \qquad (3.97) \end{aligned}$$

where (3.96) is due to Lemma 3.2. Applying Theorem 3.1 and (3.97), we have

$$p_c(\tau) = \sum_{n=0}^{M-1} \bar{p}_n = \sum_{n=0}^{M-1} \frac{1}{n!} \bar{P}^{(n)}(z)\Big|_{z=0} = \sum_{n=0}^{M-1} \frac{1}{n!}\frac{\mathrm{d}^n}{\mathrm{d}z^n}\frac{1}{C(z)}\Bigg|_{z=0}. \tag{3.98}$$

Based on Lemma 3.3 and what we exploited in the proof of Theorem 3.2, the first M coefficients of the power series $\frac{1}{C(z)}$ form the first column of the matrix inversion $\mathbf{C}_M^{-1}$, and their sum is the ℓ_1-induced matrix norm of $\mathbf{C}_M^{-1}$ as given in (3.87).

Remark 3.4 This result expresses the coverage probability of cellular networks in a very compact form, where only an inverse of a lower triangular Toeplitz matrix is needed. There exist many fast algorithms to calculate this inverse [43], which makes (3.87) more efficient than existing analytical results, e.g., [20, 28]. In addition, the class of models for which this result applies is also more general.

While general in the interferers' power gain, Proposition 3.1 loses some tractability due to the expectation over g when calculation $\{c_n\}_{n=0}^{M-1}$. The following corollary presents a more tractable expression for a specific distribution for the interferers' power gain, i.e., $g \sim \text{Gamma}(\kappa, \beta)$. Note that this is a commonly encountered distribution for the interferers' power gain in multi-antenna networks, as previously shown in [1, 8, 16, 20, 22, 25, 28, 37–39].

Corollary 3.1 *When the interferers' power gain is gamma distributed as* $g \sim \text{Gamma}(\kappa, \beta)$*, the SIR coverage probability of cellular networks is given by*

$$p_c(\tau) = \left\| \mathbf{C}_M^{-1} \right\|_1, \tag{3.99}$$

with the nonzero entries in $\mathbf{C}_M$ *as*

$$c_n = \frac{\Gamma(\kappa + n)}{\Gamma(\kappa)\Gamma(n+1)} \frac{\delta}{\delta - n} \left(\frac{\tau\beta}{\theta}\right)^n {}_2F_1\left(n + \kappa, n - \delta; n + 1 - \delta; -\frac{\tau\beta}{\theta}\right), \tag{3.100}$$

for $0 \leq n \leq M - 1$.

Proof See Appendix.

Remark 3.5 When $M = \theta = \kappa = \beta = 1$, Corollary 3.1 reduces to

$$p_c(\tau) = \frac{1}{c_0}. \tag{3.101}$$

According to (3.100), the only coefficient c_0 is simplified to

$$c_0 = {}_2F_1\,(1, -\delta; 1 - \delta; -\tau) \equiv 1 + c. \tag{3.102}$$

Note that $c = \frac{\delta\tau}{1-\delta} {}_2F_1\,(1, 1 - \delta; 2 - \delta; -\tau)$ is given by (2.63) in Sect. 2.4.1, which is the result for the SIR coverage of the single-antenna case with Rayleigh fading.

Figure 3.1 plots the SIR coverage probability of cellular networks with (3.99). In addition, our analytical results are shown to be accurate even if noise is included, which verifies the interference dominance assumption.

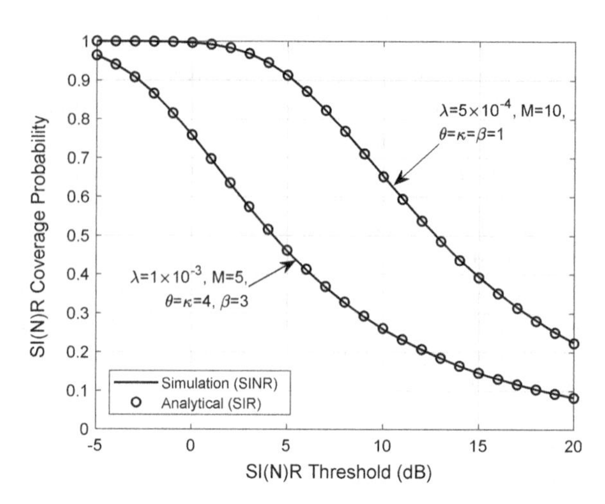

Fig. 3.1 The SI(N)R coverage probability of cellular networks when $\alpha = 4$ and $\sigma_n^2 = -97.5\,\text{dBm}$. ©2018 IEEE. Reprinted, with permission, from [13]

3.4.2 *Ad Hoc Networks*

As illustrated in Sect. 2.1.4, there is no need to calculate the integral over r_0 in (3.74) in ad hoc networks, and the resulting SIR is

$$\text{SIR} = \frac{g_{x_0} r_0^{-\alpha}}{\sum_{x \in \Phi} g_x \|x\|^{-\alpha}}, \tag{3.103}$$

where the signal power gain g_{x_0} is gamma distributed as Gamma(M, θ). Correspondingly, the coverage probability in ad hoc networks is given by the following proposition.

Proposition 3.2 *The SIR coverage probability of ad hoc networks is given by*[6]

$$p_c(\tau) = \left\| e^{\mathbf{A}_M} \right\|_1, \tag{3.104}$$

where $\mathbf{A}_M$ *is the lower triangular Toeplitz matrix with the nonzero entries as*

$$a_n = -\frac{(-1)^n}{n!} (\delta)_n \pi \lambda r_0^2 \Gamma(1-\delta) \left(\frac{\tau}{\theta}\right)^{\delta} \mathbb{E}_g\left[g^{\delta}\right], \tag{3.105}$$

for $0 \le n \le M-1$, *where the falling factorial of* x *is symbolized as* $(x)_n$.

Proof See Appendix.

[6]The matrix $\mathbf{A}_M$ has the same expression as $\mathbf{T}_M$ in (3.74). The change of notation here is mainly to distinguish the results in ad hoc networks from those under general network settings.

Remark 3.6 In the ad hoc network model, even if the noise is included, it is still feasible to derive a closed-form expression for the coverage probability. Specifically, the log-Laplace transform is given by

$$\eta(s) = -s\sigma_{\mathrm{n}}^2 - \pi\lambda\Gamma(1-\delta)s^{\delta}\mathbb{E}_g\left[g^{\delta}\right]. \tag{3.106}$$

Hence, the nonzero entries $\{a_n\}_{n=0}^{M-1}$ in (3.105) are

$$\begin{aligned} a_n &= \frac{(-s)^n}{n!}\eta^{(n)}(s) \\ &= \frac{(-1)^n}{n!}\left\{-\mathbb{1}(n\le 1)\frac{\tau r_0^{\alpha}}{\theta}\sigma_{\mathrm{n}}^2 - \pi\lambda r_0^2\Gamma(1-\delta)(\delta)_n\left(\frac{\tau}{\theta}\right)^{\delta}\mathbb{E}_g\left[g^{\delta}\right]\right\}. \end{aligned} \tag{3.107}$$

Proposition 3.2 expresses the coverage probability of ad hoc networks by an ℓ_1-induced norm of a matrix exponential. In particular, once the distribution of the interferers' power gain g is given, the nonzero entries $\{a_n\}_{n=0}^{M-1}$ in the lower triangular matrix $\mathbf{A}_M$ can be obtained according to (3.105) or (3.107). Finally, a matrix exponential is the only operation needed in the calculation. Efficient techniques exist for computing the matrix exponential of lower triangular Toeplitz matrices [44]. Similar to cellular networks, next we present a special case with closed-form expressions where the interferers' power gain is gamma distributed as $g \sim \mathrm{Gamma}(\kappa,\beta)$.

Corollary 3.2 *When the interferers' power gain is gamma distributed as* $g \sim \mathrm{Gamma}(\kappa,\beta)$*, the SINR coverage probability of ad hoc networks is given by*

$$p_{\mathrm{c}}(\tau) = \left\|e^{\mathbf{A}_M}\right\|_1, \tag{3.108}$$

with the nonzero entries in $\mathbf{A}_M$ *as*

$$a_n = \frac{(-1)^n}{n!}\left\{-\mathbb{1}(n\le 1)\frac{\tau r_0^{\alpha}}{\theta}\sigma_{\mathrm{n}}^2 - \pi\lambda r_0^2\left(\frac{\tau\beta}{\theta}\right)^{\delta}\frac{\Gamma(\delta+\kappa)\Gamma(1-\delta)\Gamma(1+\delta)}{\Gamma(\kappa)\Gamma(\delta+1-n)}\right\}, \tag{3.109}$$

for $0 \le n \le M-1$.

Proof The result follows by inserting $\mathbb{E}\left[g^{\delta}\right] = \beta^{\delta}\Gamma(\delta+\kappa)/\Gamma(\kappa)$ in (3.107).

Remark 3.7 When $M = \theta = \kappa = \beta = 1$, Corollary 3.2 reduces to

$$\begin{aligned} p_{\mathrm{c}}(\tau) &= \exp\left\{-\tau r_0^{\alpha}\sigma_{\mathrm{n}}^2 - \pi\lambda r_0^2\tau^{\delta}\Gamma(1+\delta)\Gamma(1-\delta)\right\} \\ &= \exp\left\{-\tau r_0^{\alpha}\sigma_{\mathrm{n}}^2 - \pi\lambda r_0^2\tau^{\delta}\frac{\pi\delta}{\sin\pi\delta}\right\}, \end{aligned} \tag{3.110}$$

which is exactly the same as the result for the SIR coverage of the single-antenna case with Rayleigh fading in Sect. 2.4.2.

3.5 Properties of General Multi-Antenna Networks

In the previous section, the ℓ_1-Toeplitz matrix representation in Theorem 3.2 has been applied to derive tractable expressions for the coverage in cellular and ad hoc networks. In this section, the finite sum representation in Theorem 3.1, assisted by Theorem 3.2, is applied to reveal unique properties in both types of networks, which cannot be unraveled by other analytical approaches. We investigate the effects of the transmitter density and the transmitter antenna size on the coverage probability as examples. The analysis in this part is highly nontrivial, and the clean representations in Theorems 3.1 and 3.2 make it possible.

3.5.1 *Densification: Impact of BS Density*

In [2, 41], it has been shown that for Rayleigh fading channels, i.e., $g_x \sim \text{Exp}(1)$, the SIR coverage probability is invariant to the BS density λ. In this part, we shall extend this conclusion to more general cases. Specifically, we show that, with arbitrarily distributed interferer's power gain g_x, the SIR invariance property in cellular networks still holds.

Lemma 3.4 *The SIR coverage probability in cellular networks is invariant to the BS density λ if the nearest-BS association is adopted.*

Proof The proof can be directly obtained from Corollary 3.1.

On the other hand, in ad hoc networks, since the distance between the typical receiver and the associated transmitter is fixed, the coverage probability monotonically decreases when the transmitter density increases, as the densification implies more interferers per unit area. However, there are no existing works that quantified such effect, which is pursued in the following result.

Corollary 3.3 *The SIR coverage probability* (3.104) *is a monotonically decreasing convex function of the transmitter density, and it can be rewritten as*

$$p_c(\lambda) = e^{a_0'\lambda} \sum_{n=0}^{M-1} \beta_n \lambda^n, \tag{3.111}$$

where

$$\beta_n = \frac{\left\| \left(\mathbf{A}_M' - a_0'\mathbf{I}_M\right)^n \right\|_1}{n!}, \tag{3.112}$$

and

$$a_n' = \frac{a_n}{\lambda} = -\frac{(-1)^n}{n!}(\delta)_n \pi r_0^2 \Gamma(1-\delta)\left(\frac{\tau}{\theta}\right)^\delta \mathbb{E}_g\left[g^\delta\right]. \tag{3.113}$$

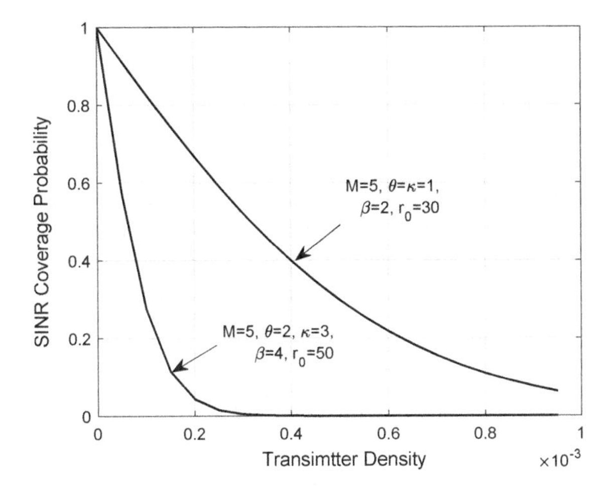

Fig. 3.2 The impact of the transmitter density on the SIR coverage probability in ad hoc networks when $\alpha = 4$ and $\tau = 0$ dB, according to (3.111). ©2018 IEEE. Reprinted, with permission, from [13]

Correspondingly, the derivative of the coverage probability with respect to the transmitter density is given by

$$\frac{\partial}{\partial \lambda} p_c(\lambda) = e^{a_0' \lambda} \left\{ a_0' \beta_{M-1} \lambda^{M-1} + \sum_{n=0}^{M-2} \left[a_0' \beta_n + (n+1)\beta_{n+1} \right] \lambda^n \right\}. \tag{3.114}$$

Proof See Appendix.

From Corollary 3.3 we have $p_c(\lambda) \to 1$ as $\lambda \to 0$, which is independent of all the other network parameters. Hence, for any coverage requirement $1 - \varepsilon$ at the typical receiver, there exists a maximum transmitter density λ that can satisfy it regardless of the other network parameters, and this density can be numerically determined. Furthermore, this result fully characterizes how the transmitter density affects the coverage probability, which is shown in Fig. 3.2. In particular, we prove that increasing the transmitter density degrades the coverage probability in ad hoc networks, and the coverage probability is a product of an exponential function and a polynomial function of order $M - 1$ of the transmitter density λ. For the special case of $M = 1$, i.e., single-antenna networks with Rayleigh fading channel, the coverage probability reduces to an exponential one. In other words, the multi-antenna setting increases the coverage probability by the additional polynomial term. In addition, the derivative given in (3.114) reflects the sensitivity of the coverage probability with respect to the transmitter density.

Remark 3.8 A related result on the impact of the bipolar distance r_0 can be readily obtained from Corollary 3.3. Since r_0^2 and λ are interchangeable in (3.111), there exists a duality between λ and r_0^{-2}, where the former affects the interference power while the latter only affects the signal power. The impact of the bipolar distance r_0 is then given by

$$p_c(r_0) = e^{\hat{a}_0 r_0^2} \sum_{n=0}^{M-1} \hat{\beta}_n r_0^{2n}, \tag{3.115}$$

where $\hat{\beta}_n = \frac{\left\| \left(\hat{\mathbf{A}}_M - \hat{a}_0 \mathbf{I}_M \right)^n \right\|_1}{n!}$ and $\hat{a}_n = a_n / r_0^2$. Similar to Corollary 3.3, it can be proved that the coverage probability is a monotonically decreasing convex function of the bipolar distance. In addition, the monotonicity and convexity in Corollary 3.3 are also applicable to the SINR coverage probability where the noise is also taken into consideration, and the proof can be found in Appendix.

3.5.2 *MIMO: Impact of Antenna Size*

As discussed in Sect. 3.3.4, the framework generalizes our ability in analyzing the single-antenna network to the multi-antenna one. Hence, it is intriguing to apply it to investigate how multi-antenna techniques affect the coverage probability. In the following, we shall perform such an investigation by taking a multiple-input single-output (MISO) network with MRT beamforming as an example. In this case, the signal power gain g_{x_0} is gamma distributed as Gamma(M, θ), where M is the number of transmit antennas.

Remark 3.9 As shown in Table 3.1, the number of antennas is typically related to the shape parameter M in the gamma distribution of the signal power gain g_{x_0}. Hence, the derivations and conclusions in the following are also applicable to other network parameters related to the shape parameter M, e.g., the user number U_k in MIMO HetNets [3], the number of coordination requests K_{x_0} in user-centric interference coordination [15], and the number of transmitted streams N_{x_0} in physical layer security-aware networks [17].

We first present a general lemma that will be used in the following derivation. We define the *coverage improvement* for the n-th antenna as the increment of the coverage probability when the antenna size is enlarged from $n-1$ to n.

Lemma 3.5 *For both ad hoc and cellular networks, the coverage improvement due to the $M+1$-th antenna is*

$$p_c(M+1) - p_c(M) = \bar{p}_M. \tag{3.116}$$

For ad hoc networks, $\bar{p}_n = p_n$ while for cellular $\bar{p}_n = \mathbb{E}_{r_0}[p_n]$, with $\{p_n\}_{n=0}^{\infty}$ given in Lemma 3.1.

Proof The result follows directly from Theorem 3.1.

Intuitively, enlarging the antenna size increases both the information signal power and the interference power, hence an explicit analysis is needed to reveal the overall effect. Based on Lemma 3.5, we have the following result.

Proposition 3.3 *For both ad hoc and cellular networks, increasing the antenna size always improves the coverage probability, i.e.,* $\bar{p}_n > 0$ *for* $n > 0$.

Proof According to (3.74) and (3.136), we have

$$\begin{aligned} p_c &= \mathbb{E}_{r_0}\left[\left\|e^{\mathbf{T}_M}\right\|_1\right] \\ &= \mathbb{E}_{r_0}\left[e^{t_0}\left(1+\sum_{n=1}^{M-1}\frac{\left\|(\mathbf{T}_M - t_0\mathbf{I}_M)^n\right\|_1}{n!}\right)\right]. \end{aligned} \tag{3.117}$$

Hence, $\bar{p}_n$ can be rewritten as

$$\bar{p}_n = \mathbb{E}_{r_0}\left[e^{t_0}\frac{\left\|(\mathbf{T}_M - t_0\mathbf{I}_M)^n\right\|_1}{n!}\right]. \tag{3.118}$$

Similar to (3.133), it can be proved that $t_0 < 0$ while $t_n > 0$ for $n > 0$. In this way, all the entries in the strict lower triangular matrix $\mathbf{T}_M - t_0\mathbf{I}_M$ are nonnegative, and so are $\{\bar{p}_n\}_{n=0}^{\infty}$.

Note that Proposition 3.3 applies to very general network settings, as long as the signal power gain is gamma distributed, and the assumption that the shape parameter M is the only parameter related to the number of antennas. In the following, we apply this result to different network models.

Proposition 3.4 *Denoting the outage probability in multi-antenna cellular networks by* $p_o(M)$, *we have*

$$\lim_{M\to\infty}\frac{p_o(M)}{p_o(M+1)} = \lim_{n\to\infty}\frac{\bar{p}_n}{\bar{p}_{n+1}} = r_c > 1, \tag{3.119}$$

where r_c *is the radius of convergence of the power series* $\bar{P}(z)$ *in (3.97), given by the solution to the equation*

$$\mathbb{E}_g\left[{}_1F_1\left(-\delta; 1-\delta; \frac{(r_c-1)\tau}{\theta}g\right)\right] = 0. \tag{3.120}$$

Proof See Appendix.

A corollary of this result is given next when the interferers' power gain is gamma distributed as $g \sim \text{Gamma}(\kappa, \beta)$.

Corollary 3.4 *When the interferers' power gain in multi-antenna cellular networks is gamma distributed, we have*

$$\lim_{M\to\infty}\frac{p_o(M)}{p_o(M+1)} = r_c, \tag{3.121}$$

where $r_c \in \left(1, 1+\frac{\theta}{\tau\beta}\right)$ *is the solution to the equation*

$$ {}_2F_1\left(\kappa, -\delta; 1-\delta, (r_c - 1)\frac{\tau\beta}{\theta}\right) = 0. \tag{3.122}$$

Proof It is proved by plugging the pdf of g, i.e., $f_g(u) = \frac{u^{\kappa-1}e^{-\frac{u}{\beta}}}{\beta^\kappa \Gamma(\kappa)}$, into (3.120), and the upper bound of r_c can be easily obtained from the radius of convergence of the Gaussian hypergeometric function.

Remark 3.10 Proposition 3.4 indicates that, when M is large, the coverage improvement of adding the n-th antenna is r_c times larger than that of adding the $(n+1)$-th antenna. Furthermore, the outage probability of cellular networks in the logarithmic scale decreases linearly in M with slope $-\log_{10} r_c$.

Figure 3.3 shows the SIR outage probability of cellular networks versus the antenna size. While Proposition 3.4 is an asymptotic result, it is quite accurate also when the number of antennas is small. In addition, as r_c is larger than 1 in Proposition 3.4, it demonstrates that increasing the antenna size definitely benefits the coverage probability, and it also shows that the coverage improvement $\bar{p}_n$ diminishes as the number of antennas grows large. However, this may not be the case in ad hoc networks, as shown next.

Analyzing the coverage improvement in ad hoc networks for general network settings is more challenging, so we start from the special case $\alpha = 4$, which is usually used in existing works [2, 26] for analytical tractability. Particularly, we focus on finding the antenna index, denoted by $n^\star + 1$, that contributes the most significant coverage improvement in ad hoc networks.

Proposition 3.5 *When the path loss exponent* $\alpha = 4$*, the SIR coverage improvement due to adding the* $n+1$*-th antenna in ad hoc multi-antenna networks monotonically*

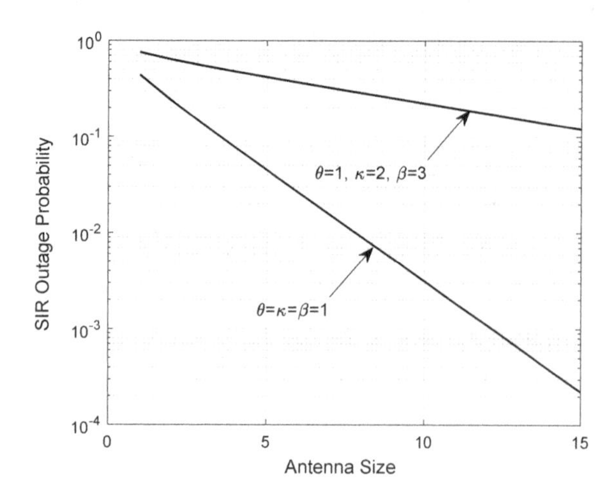

Fig. 3.3 The SIR outage probability of cellular networks when $\tau = 0$ dB and $\alpha = 4$, according to (3.99). ©2018 IEEE. Reprinted, with permission, from [13]

decreases in the interval

$$n > \frac{\mu^2}{4} - 1, \tag{3.123}$$

where $\mu > 0$ *is given by*

$$\mu = \pi \lambda r_0^2 \Gamma(1-\delta) \left(\frac{\tau}{\theta}\right)^{\delta} \mathbb{E}_g \left[g^{\delta}\right]. \tag{3.124}$$

Proof See Appendix.

Proposition 3.5 indicates that the largest coverage improvement occurs when adding one of the first $\left\lceil \frac{\mu^2}{4} - 1 \right\rceil + 1$ antennas, i.e., $1 \leq n^\star \leq \left\lceil \frac{\mu^2}{4} - 1 \right\rceil$. Furthermore, the condition that the coverage improvement is always monotonically decreasing can be derived via Proposition 3.5, given by $\frac{\mu^2}{4} - 1 < 0$, i.e., $\mu < 2$.

The SIR coverage improvement of ad hoc networks when $\alpha = 4$ is presented in Fig. 3.4a. The situations when the coverage improvement has a peak value, i.e., $\mu > 2$, are of particular interest. It can be discovered that the denser the network (or, equivalently, the longer the bipolar distance), the larger the index of the antenna that provides the maximum coverage improvement. Note that we exploit an upper bound in (3.146), and therefore, $\left\lceil \frac{\mu^2}{4} - 1 \right\rceil + 1$ is an upper bound for the antenna index $n^\star + 1$ with the most significant contribution in terms of the coverage improvement. In Fig. 3.4a, we see that this upper bound is very tight, which demonstrates the effectiveness of the result in Proposition 3.5 and the analytical framework.

For the general case $\alpha > 2$, although it is difficult to obtain similar analytical results as Proposition 3.5 on the monotonicity of the coverage improvement, a closed-form expression for the coverage improvement is given in the following proposition, which can be used to numerically test the monotonicity property.

Proposition 3.6 *The SIR coverage improvement of the* $n+1$*-th antenna in ad hoc networks is given by*

$$\bar{p}_n = \frac{(-1)^n e^{-\mu}}{n!} \sum_{k=1}^{n} S(n,k) T_k(-\mu) \delta^k, \tag{3.125}$$

where $S(n,k)$ *are the Stirling numbers of the first kind, and* $T_k(x)$ *denotes the Touchard polynomial [23].*

Proof See Appendix.

Remark 3.11 The Touchard polynomial of order n is obtained when calculating the n-th moment of a Poisson distributed random variable. The appearance of such polynomial in (3.125) is related to the falling factorial and the Taylor expansion of the exponential function.

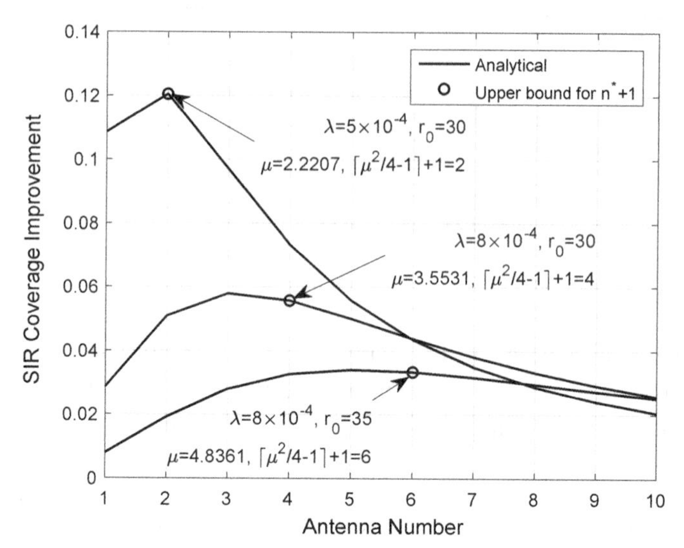

(a) The increment of the SIR coverage probability of ad hoc networks when SNR $= 0$ dB, $\theta = \kappa = \beta = 1$, and $\alpha = 4$. © 2018 IEEE. Reprinted, with permission, from [13].

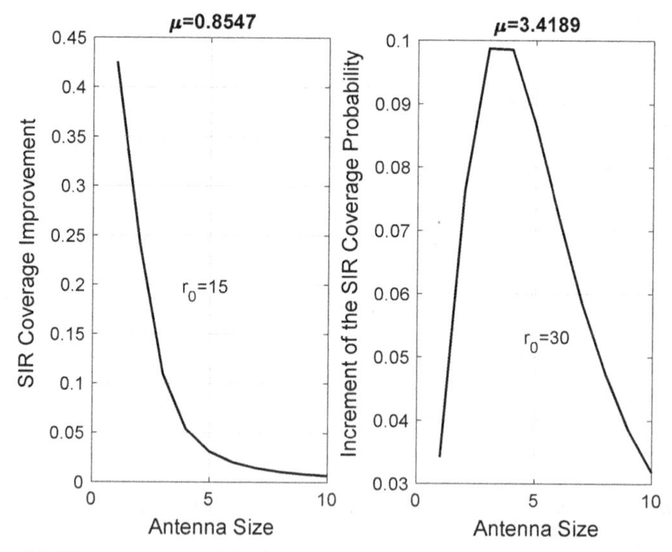

(b) The increment of the SIR coverage probability (3.125) of ad hoc networks when SNR $= 0$ dB, $\lambda = 5 \times 10^{-3}$, $\theta = \kappa = \beta = 1$, and $\alpha = 3$. © 2018 IEEE. Reprinted, with permission, from [13].

Fig. 3.4 The coverage improvement in ad hoc networks. When $1 - \mu\delta > 0$, the coverage improvement monotonically decreases with the antenna index, while there exists a peak value of the coverage improvement when $1 - \mu\delta \le 0$

Figure 3.4b plots the SIR coverage improvement as the number of antennas increases. It is numerically found that the coverage improvement has two totally different behaviors when the number of antennas increases:

- When $p_0 > p_1$, i.e., $1 - \mu\delta > 0$, the coverage improvement is monotonically decreasing with the antenna size, which is similar to that in cellular networks. In other words, the coverage improvement would never increase once it decreases at the beginning.
- When $1 - \mu\delta \leq 0$, the coverage improvement has a peak value $p_{n^\star}$ when the antenna size is enlarged, and the optimal value $n^\star$ can be numerically determined by the closed forms in Proposition 3.6. This means that adding the $(n^\star + 1)$-th antenna is the most effective in terms of the coverage improvement. Note that, for the special case that $\alpha = 4$, we have derived in Proposition 3.5 that the coverage improvement monotonically decreases when $\mu < 2$. This is a special case of the condition $1 - \mu\delta > 0$ when $\delta = \frac{1}{2}$, which verifies both the effectiveness of the analytical results in Proposition 3.5 and the reasoning of the conclusion drawn from the simulations results for general cases.

3.6 Summary

This chapter presented a unified analytical framework for coverage analysis of general multi-antenna wireless networks. Various tractable analytical results for the coverage probability were demonstrated. In particular, expressions for a general network model was first derived. Two typical network models, i.e., cellular and ad hoc networks, were then investigated to demonstrate the generality and effectiveness of the analytical framework. More importantly, system insights, i.e., the impacts of the transmitter density and the antenna size, were analytically revealed via the framework in different multi-antenna networks. Overall, this chapter provides a powerful toolbox for the evaluation and design of various multi-antenna wireless networks, which shall be applied to investigate different network models in the following chapters. In particular, refined analysis of various network performance metrics in typical network models will be presented in Chaps. 4 and 5 will apply the analytical framework for network performance optimization.

Bibliographical Notes

The performance analysis of various MIMO networks via stochastic geometry has been conducted since 2008. Chandrasekhar et al. (2009) analyzed the coverage probability of two-tier MIMO HetNets, and the transmission capacity of multi-antenna ad hoc networks was derived by Hunter et al. (2008). Both of them applied the first-order Taylor expansion to provide approximations for different performance metrics. Later on, different inequalities, e.g., Markov's inequality in Jindal et al. (2011), Chebyshev's inequality in Huang et al. (2012), the union bound in Dhillon et al. (2013), and an upper bound for the cumulative distribution function of gamma random variables in Bai and Heathh Jr. (2015), have been adopted to make the analysis tractable.

On the other hand, exact analytical evaluations have also been investigated. The rate coverage probability was derived in Wu et al. (2015), while Shojaeifard et al. (2016) investigated the downlink spectral efficiency. Both of these two studies are with improper integrals that are inefficient for numerical evaluation. In addition, closed-form expressions for coverage probabilities were obtained in Louie et al. (2011) for ad hoc networks, Zhang et al. (2013) for security-aware networks, and Gupta et al. (2014) for MIMO HetNets, and the results were further generalized by Wu et al. (2012). More recently, Li et al. (2014, 2015, 2016) presented tractable analysis for ASE and energy efficiency in small cell networks, efficient design for user-centric interference management, and the fundamental trade-off between ASE and link reliability. Yu et al. (2018) extended these works to a general analytical framework for multi-antenna wireless networks.

Appendix

Proof of Corollary 3.1

Since $g \sim \text{Gamma}(\kappa, \beta)$, i.e., $f_g(u) = \frac{u^{\kappa-1}e^{-\frac{u}{\beta}}}{\beta^\kappa \Gamma(\kappa)}$, according to (3.88), (3.89), and (3.92), we have

$$c_n = \frac{(-s)^n}{n!}\frac{\mathrm{d}^n}{\mathrm{d}s^n}\left[1 - \frac{\eta(s)}{\pi\lambda r_0^2}\right] \tag{3.126}$$

$$= -\frac{(-s)^n}{n!}\frac{\mathrm{d}^n}{\mathrm{d}s^n}\left\{\delta(sr_0^{-\alpha})^\delta \mathbb{E}_g\left[g^\delta \gamma(-\delta, sr_0^{-\alpha}g)\right]\right\}$$

$$= -\frac{(-s)^n}{n!}\frac{\mathrm{d}^n}{\mathrm{d}s^n}\int_1^\infty \mathbb{E}_g\left[\exp\left(-sr_0^{-\alpha}v^{-\frac{\alpha}{2}}g\right)\right]\mathrm{d}v \tag{3.127}$$

$$= -\frac{(-s)^n}{n!}\int_1^\infty \left[\frac{\mathrm{d}^n}{s^n}\frac{1}{\left(1+\beta r_0^{-\alpha}v^{-\frac{\alpha}{2}}s\right)^\kappa}\right]\mathrm{d}v$$

$$= -\frac{\Gamma(\kappa+n)}{\Gamma(\kappa)\Gamma(n+1)}\left(\frac{\tau\beta}{\theta}\right)^{\frac{2}{\alpha}}\int_{\left(\frac{\tau\beta}{\theta}\right)^{-\frac{2}{\alpha}}}^\infty \frac{\left(v^{-\frac{\alpha}{2}}\right)^n}{\left(1+v^{-\frac{\alpha}{2}}\right)^{\kappa+n}}\mathrm{d}v$$

$$= \frac{\Gamma(\kappa+n)}{\Gamma(\kappa)\Gamma(n+1)}\frac{\delta}{\delta-n}\left(\frac{\tau\beta}{\theta}\right)^n {}_2F_1\left(n+\kappa, n-\delta; n+1-\delta; -\frac{\tau\beta}{\theta}\right),$$

where (3.127) follows from the definition of the lower incomplete gamma function $\gamma(s, x)$, and the last equality follows from the integral representation of the hypergeometric function [23, Sect. 9.14], which completes the proof.

Proof of Proposition 3.2

According to the two steps of applying Theorem 3.2 presented in Remark 4, first, we calculate the conditional Laplace transform, expressed as

$$\mathscr{L}(s) = \exp\left\{-2\pi\lambda \int_0^{\infty} \left(1 - \mathbb{E}_g[\exp(-sgv^{-\alpha})]\right)v\mathrm{d}v\right\}. \tag{3.128}$$

To obtain a coverage probability expression for arbitrarily distributed interferers' power gains, we propose to swap the order of the integral and the expectation. In this way, part of the exponent is given by

$$\begin{aligned}
&2\mathbb{E}_g\left\{\int_0^{\infty} \left[1 - \exp(-sgv^{-\alpha})\right]v\mathrm{d}v\right\} \\
&= \mathbb{E}_g\left\{(sg)^{\frac{2}{\alpha}} \frac{2}{\alpha}\int_0^1 \frac{v}{1-v}\left[-\ln(1-v)\right]^{-\frac{2}{\alpha}-1}\mathrm{d}v\right\} \\
&= \mathbb{E}_g\left\{(sg)^{\delta}\Gamma(1-\delta)\right\}.
\end{aligned} \tag{3.129}$$

Therefore, the log-Laplace transform can be written as

$$\eta(s) = -\pi\lambda\Gamma(1-\delta)s^{\delta}\mathbb{E}_g\left[g^{\delta}\right], \tag{3.130}$$

and the nonzero entries of $\mathbf{A}_M$ are determined by

$$\begin{aligned}
a_n &= \frac{(-s)^n}{n!}\eta^{(n)}(s) \\
&= -\frac{(-1)^n}{n!}\pi\lambda r_0^2\Gamma(1-\delta)(\delta)_n\left(\frac{\tau}{\theta}\right)^{\delta}\mathbb{E}_g\left[g^{\delta}\right].
\end{aligned} \tag{3.131}$$

Since there is no need to take an expectation over r_0 in the ad hoc network model, the derivation steps similar to (3.97) and (3.98) are unnecessary, and the proof is complete.

Proof of Corollary 3.3

According to (3.106), the Laplace transform of noise and interference is

$$\mathscr{L}(s) = p_0 = e^{\eta(s)} = \exp\left(-s\sigma_{\mathrm{n}}^2 - \pi\lambda\Gamma(1-\delta)s^{\delta}\mathbb{E}_g\left[g^{\delta}\right]\right). \tag{3.132}$$

Note that $\Gamma(1-\delta)$ is a positive term due to the fact that $0 < \delta < 1$. Hence, the Laplace transform p_0 is a convex and monotonically decreasing function with respect to the transmitter density λ.

Furthermore, according to (3.107), the signs of $\{a_n\}_{n=1}^{M-1}$ are critical, i.e.,

$$a_n = -\frac{(-1)^n}{n!}(\delta)_n\pi\lambda\Gamma(1-\delta)s^{\delta}\mathbb{E}_g\left[g^{\delta}\right] + s\sigma_{\mathrm{n}}^2\mathbb{1}(n=1). \tag{3.133}$$

Since $(-1)^n(\delta)_n = (-\delta)^{(n)} < 0$ with $(x)^{(n)}$ denoting the rising factorial, we have $a_n > 0$ for $1 \leq n \leq M$. Recall that the recursive relations between $\{p_n\}_{n=1}^{M-1}$ are

$$p_n = \sum_{i=0}^{n-1} \frac{n-i}{n} a_{n-i} p_i. \tag{3.134}$$

Since the term $\frac{n-i}{n} a_{n-i}$ is positive, it turns out that all $\{p_n\}_{n=1}^{M-1}$ have the same monotonicity and convexity with respect to λ. Recalling that $p_c(\tau) = \sum_{n=0}^{M-1} p_n$, the monotonicity and concavity in Corollary 4.3 have been proved. Next, we prove the expression (3.111).

We first write $\mathbf{A}'_M$ in the form

$$\mathbf{A}'_M = a'_0 \mathbf{I}_M + (\mathbf{A}'_M - a'_0 \mathbf{I}_M). \tag{3.135}$$

Since $\mathbf{A}'_M$ is a lower triangular Toeplitz matrix, the second part is a nilpotent matrix, i.e., $(\mathbf{A}'_M - a'_0 \mathbf{I}_M)^n = \mathbf{0}$ for $n \geq M$. Hence, according to the properties of matrix exponential, we have

$$e^{\mathbf{A}_M} = e^{\lambda \mathbf{A}'_M} = e^{a'_0 \lambda} \cdot \sum_{n=0}^{M-1} \frac{1}{n!} \left[\lambda \left(\mathbf{A}'_M - a'_0 \mathbf{I}_M\right)\right]^n. \tag{3.136}$$

Since it has been shown that $a'_n > 0$ for $n \geq 1$, $\mathbf{A}'_M - a'_0 \mathbf{I}_M$ is a strictly lower triangular Toeplitz matrix with all positive entries, and so are the matrices $(\mathbf{A}'_M - a'_0 \mathbf{I}_M)^n$. Therefore,

$$\left\| e^{\lambda \mathbf{A}'_M} \right\|_1 = e^{a'_0 \lambda} \cdot \sum_{n=0}^{M-1} \frac{1}{n!} \left[\lambda^n \left\| \left(\mathbf{A}'_M - a'_0 \mathbf{I}_M\right)^n \right\|_1\right], \tag{3.137}$$

which completes the proof of Corollary 4.3.

Proof of Proposition 3.4

According to Theorem 3.1, the outage probability is $p_o(\tau) = 1 - \sum_{n=0}^{M-1} \bar{p}_n$, and then we have

$$\begin{aligned} \lim_{M \to \infty} \frac{p_o(M+1)}{p_o(M)} &= 1 - \lim_{M \to \infty} \frac{\bar{p}_M}{1 - \sum_{n=0}^{M-1} \bar{p}_n} \\ &= 1 - \lim_{M \to \infty} \frac{1}{1 - \sum_{n=M}^{\infty} \frac{\bar{p}_n}{\bar{p}_M}}. \end{aligned} \tag{3.138}$$

Since r_c is the radius of convergence of the power series $\bar{P}(z)$, i.e., $r_c = \lim_{n \to \infty} \frac{\bar{p}_n}{\bar{p}_{n+1}}$, the above equation can be further simplified as

$$\lim_{M\to\infty}\frac{p_{\mathrm{o}}(M+1)}{p_{\mathrm{o}}(M)}=1-\lim_{M\to\infty}\frac{1}{\sum_{n=0}^{\infty}\left(\frac{1}{r_{\mathrm{c}}}\right)^{n}}=\frac{1}{r_{\mathrm{c}}}. \tag{3.139}$$

According to (3.126), the coefficients in the power series $C(z)$ are given by

$$c_n=\frac{(-s)^n}{n!}c_0^{(n)}(s), \tag{3.140}$$

where $c_0(s)=-\delta(sr_0^{-\alpha})^{\delta}\mathbb{E}_g\left[g^{\delta}\gamma(-\delta,sr_0^{-\alpha}g)\right]$. By reversely applying the Taylor expansion, the power series $C(z)$ can be written as

$$C(z)=\sum_{n=0}^{\infty}c_nz^n=\sum_{n=0}^{\infty}\frac{(-sz)^n}{n!}c_0^{(n)}(s)=c_0((1-z)s). \tag{3.141}$$

Recalling that in (3.97) we proved that $\bar{P}(z)=\frac{1}{C(z)}$, thus the radius of convergence of $\bar{P}(z)$ is the solution of the equation $C(r_{\mathrm{c}})=c_0((1-r_{\mathrm{c}})s)=0$, which is equivalent to (3.120).

Next, we prove that the solution r_{c} to equation (3.120) is larger than 1. The left-hand side of (3.120) can be rewritten as

$$\mathbb{E}_g\left[{}_1F_1\left(-\delta;1-\delta;\frac{(r_{\mathrm{c}}-1)\tau}{\theta}g\right)\right]1+\delta\mathbb{E}_g\left[\int_0^1\frac{1-e^{\frac{(r_{\mathrm{c}}-1)\tau}{\theta}gv}}{v^{1+\delta}}\mathrm{d}v\right]. \tag{3.142}$$

Since $0<\delta<1$, $\tau>0$, $\beta>0$, and g is assumed as a nonnegative random variable with arbitrary distributions, it is seen from (3.142) that $C(r_{\mathrm{c}})$ is a monotonically decreasing function of r_{c}. Furthermore, it is easy to check that, when $r_{\mathrm{c}}=1$, we have $C(1)=1$. Following the monotonicity of $C(r_{\mathrm{c}})$ and the fact that $C(1)>0$, we conclude that there exists only one solution of (3.120) that is larger than 1.

Proof of Proposition 3.5

According to (3.131), we have

$$\begin{aligned}A(z)&=\sum_{n=0}^{\infty}a_nz^n=\sum_{n=0}^{\infty}\frac{(-sz)^n}{n!}\eta^{(n)}(s)=\eta((1-z)s)\\&=-\pi\lambda r_0^2\Gamma(1-\delta)\mathbb{E}_g\left[g^{\delta}\right](1-z)^{\delta}=-\mu(1-z)^{\delta}.\end{aligned} \tag{3.143}$$

Therefore, with the formulas (3.95) and (3.97), we have the closed-form expression

$$\bar{P}(z)=P(z)=e^{A(z)}=e^{-\mu(1-z)^{\delta}}. \tag{3.144}$$

When $\alpha = 4$, i.e., $\delta = 1/2$, the power series $\bar{P}(z)$ is given by $\bar{P}(z) = \sum_{n=0}^{\infty} \bar{p}_n z^n = e^{-\mu\sqrt{1-z}}$. According to the definition of the modified Bessel function of the second kind $K_n(x)$ [45, Ch. 9.6], we have

$$\bar{p}_n = \sqrt{\frac{2\mu}{\pi}} \frac{(\mu/2)^n}{n!} K_{n-\frac{1}{2}}(\mu). \tag{3.145}$$

Then, define the ratio to test the monotonicity as

$$\begin{aligned}\frac{\bar{p}_{n+1}}{\bar{p}_n} &= \frac{\mu}{2(n+1)} \frac{K_{n+\frac{1}{2}}(\mu)}{K_{n-\frac{1}{2}}(\mu)} \\ &\leq \frac{n + \sqrt{n^2 + \mu^2}}{2(n+1)},\end{aligned} \tag{3.146}$$

where the inequality adopted in (3.146) comes from [46, Th. 1]. Finally, it can be checked that $\frac{n+\sqrt{n^2+\mu^2}}{2(n+1)} < 1$ when $n > \frac{\mu^2}{4} - 1$, which completes the proof.

Proof of Proposition 3.6

By performing coefficient extraction to (3.144), we have

$$\begin{aligned}\bar{P}(z) = e^{\mu(1-z)^\delta} &= \sum_{k=0}^{\infty} \frac{\mu^k (1-z)^{\delta k}}{k!} \\ &= \sum_{k=0}^{\infty} \frac{\mu^k}{k!} \sum_{n=0}^{\infty} \frac{(-1)^n}{n!} (\delta k)_n z^n \sum_{n=0}^{\infty} \left[\frac{(-1)^n}{n!} \sum_{k=0}^{\infty} \frac{\mu^k}{k!} (\delta k)_n \right] z^n,\end{aligned} \tag{3.147}$$

and it follows that

$$\begin{aligned}\bar{p}_n &= \frac{(-1)^n}{n!} \sum_{k=0}^{\infty} \frac{(-\mu)^k}{k!} (\delta k)_n \\ &= \frac{(-1)^n}{n!} \sum_{k=0}^{\infty} \frac{(-\mu)^k}{k!} \sum_{p=0}^{n} \rho(n, p)(\delta k)^p \qquad (3.148) \\ &= \frac{(-1)^n}{n!} \sum_{p=0}^{n} \rho(n, p)\delta^p \sum_{k=0}^{\infty} \frac{(-\mu)^k}{k!} k^p \\ &= \frac{(-1)^n e^{-\mu}}{n!} \sum_{k=1}^{n} \rho(n, k) T_k(-\mu) \delta^k, \qquad (3.149)\end{aligned}$$

where steps (3.148) and (3.149) reversely apply the definition of the Stirling numbers of the first kind and the Touchard polynomial, respectively.

References

1. A.M. Hunter, J.G. Andrews, S. Weber, Transmission capacity of ad hoc networks with spatial diversity. IEEE Trans. Wirel. Commun. **7**, 5058–5071 (2008)
2. J.G. Andrews, F. Baccelli, R.K. Ganti, A tractable approach to coverage and rate in cellular networks. IEEE Trans. Commun. **59**, 3122–3134 (2011)
3. C. Li, J. Zhang, J.G. Andrews, K.B. Letaief, Success probability and area spectral efficiency in multiuser MIMO HetNets. IEEE Trans. Commun. **64**, 1544–1556 (2016)
4. H.S. Jo, Y.J. Sang, P. Xia, J.G. Andrews, Heterogeneous cellular networks with flexible cell association: a comprehensive downlink SINR analysis. IEEE Trans. Wireless Commun. **11**, 3484–3495 (2012)
5. F. Rusek, D. Persson, B.K. Lau, E.G. Larsson, T.L. Marzetta, O. Edfors, F. Tufvesson, Scaling up MIMO: opportunities and challenges with very large arrays. IEEE Signal Process. Mag. **30**, 40–60 (2013)
6. G.L. Stuber, J.R. Barry, S.W. McLaughlin, Y. Li, M.A. Ingram, T.G. Pratt, Broadband MIMO-OFDM wireless communications. Proc. IEEE **92**, 271–294 (2004)
7. A. Paulraj, R. Nabar, D. Gore, *Introduction to space-time wireless communications* (Cambridge University Press, 2003)
8. R.H.Y. Louie, M.R. McKay, I.B. Collings, Open-loop spatial multiplexing and diversity communications in ad hoc networks. IEEE Trans. Inf. Theory **57**, 317–344 (2011)
9. D. Gesbert, M. Kountouris, R.W. Heath Jr., C. Chae, T. Salzer, Shifting the mimo paradigm. IEEE Signal Process. Mag. **24**, 36–46 (2007)
10. M. Costa, Writing on dirty paper (corresp.). IEEE Trans. Inf. Theor. **29**, 439–441 (1983)
11. H. Harashima, H. Miyakawa, Matched-transmission technique for channels with intersymbol interference. IEEE Trans. Commun. **20**, 774–780 (1972)
12. Q.H. Spencer, A.L. Swindlehurst, M. Haardt, Zero-forcing methods for downlink spatial multiplexing in multiuser MIMO channels. IEEE Trans. Signal Process. **52**, 461–471 (2004)
13. X. Yu, C. Li, J. Zhang, M. Haenggi, K.B. Letaief, A unified framework for the tractable analysis of multi-antenna wireless networks. IEEE Trans. Wirel. Commun. **17**, 7965–7980 (2018)
14. C. Li, J. Zhang, K.B. Letaief, Throughput and energy efficiency analysis of small cell networks with multi-antenna base stations. IEEE Trans. Wirel. Commun. **13**, 2505–2517 (2014)
15. C. Li, J. Zhang, M. Haenggi, K.B. Letaief, User-centric intercell interference nulling for downlink small cell networks. IEEE Trans. Commun. **63**, 1419–1431 (2015)
16. N. Jindal, J.G. Andrews, S. Weber, Multi-antenna communication in ad hoc networks: achieving MIMO gains with SIMO transmission. IEEE Trans. Commun. **59**, 529–540 (2011)
17. X. Yu, C. Li, J. Zhang, K.B. Letaief, A tractable framework for performance analysis of dense multi-antenna networks, in *Proceedings of IEEE International Conference on Communications (ICC)*, (Paris, France), pp. 1–6, 2017
18. X. Yu, J. Zhang, M. Haenggi, K.B. Letaief, Coverage analysis for millimeter wave networks: the impact of directional antenna arrays. IEEE J. Sel. Areas Commun. **35**, 1498–1512 (2017)
19. H. Huang, C.B. Papadias, S. Venkatesan, *MIMO communication for cellular networks.* (Springer Science & Business Media, 2011)
20. X. Zhang, X. Zhou, M.R. McKay, Enhancing secrecy with multi-antenna transmission in wireless ad hoc networks. IEEE Trans. Inf. Forensics Secur. **8**, 1802–1814 (2013)
21. R.W. Heath Jr., T. Wu, Y.H. Kwon, A.C.K. Soong, Multiuser MIMO in distributed antenna systems with out-of-cell interference. IEEE Trans. Signal Process. **59**, 4885–4899 (2011)
22. Y. Wu, R.H.Y. Louie, M.R. McKay, I.B. Collings, Generalized framework for the analysis of linear MIMO transmission schemes in decentralized wireless ad hoc networks. IEEE Trans. Wirel. Commun. **11**, 2815–2827 (2012)
23. D. Zwillinger, *Table of integrals, series, and products* (Elsevier, Amsterdam, Netherlands, 2014)
24. M. Haenggi, *Stochastic geometry for wireless networks* (Cambridge University Press, Cambridge, U.K., 2012)

25. V. Chandrasekhar, M. Kountouris, J.G. Andrews, Coverage in multi-antenna two-tier networks. IEEE Trans. Wirel. Commun. **8**, 5314–5327 (2009)
26. T. Bai, R.W. Heath Jr., Coverage and rate analysis for millimeter-wave cellular networks. IEEE Trans. Wirel. Commun. **14**, 1100–1114 (2015)
27. A. Thornburg, T. Bai, R.W. Heath Jr., Performance analysis of outdoor mmWave ad hoc networks. IEEE Trans. Signal Process. **64**, 4065–4079 (2016)
28. A.K. Gupta, H.S. Dhillon, S. Vishwanath, J.G. Andrews, Downlink multi-antenna heterogeneous cellular network with load balancing. IEEE Trans. Commun. **62**, 4052–4067 (2014)
29. S. Roman, The formula of Faà di Bruno. The Am. Math. Mon. **87**(10), 805–809 (1980)
30. S. Weber, J.G. Andrews, N. Jindal, The effect of fading, channel inversion, and threshold scheduling on Ad Hoc networks. IEEE Trans. Inf. Theor. **53**, 4127–4149 (2007)
31. G.-C. Rota, The number of partitions of a set. The Am. Math. Mon. **71**(5), 498–504 (1964)
32. A. Shojaeifard, K.A. Hamdi, E. Alsusa, D.K.C. So, J. Tang, A unified model for the design and analysis of spatially-correlated load-aware HetNets. IEEE Trans. Commun. **62**, 1–16 (2014)
33. M. Kountouris, J.G. Andrews, Downlink SDMA with limited feedback in interference-limited wireless networks. IEEE Trans. Wirel. Commun. **11**, 2730–2741 (2012)
34. C. Saha, M. Afshang, H.S. Dhillon, 3GPP-Inspired HetNet model using poisson cluster process: Sum-product functionals and downlink coverage. IEEE Trans. Commun. **66**, 2219–2234 (2018)
35. X. Zhang, J.G. Andrews, Downlink cellular network analysis with multi-slope path loss models. IEEE Trans. Commun. **63**, 1881–1894 (2015)
36. R.K. Ganti, M. Haenggi, Asymptotics and approximation of the SIR distribution in general cellular networks. IEEE Trans. Wirel. Commun. **15**, 2130–2143 (2016)
37. K. Huang, J.G. Andrews, D. Guo, R.W. Heath Jr., R.A. Berry, Spatial interference cancellation for multiantenna mobile ad hoc networks. IEEE Trans. Inf. Theor. **58**, 1660–1676 (2012)
38. A. Shojaeifard, K.A. Hamdi, E. Alsusa, D.K.C. So, J. Tang, K.K. Wong, Design, modeling, and performance analysis of multi-antenna heterogeneous cellular networks. IEEE Trans. Commun. **64**, 3104–3118 (2016)
39. Y. Wu, Y. Cui, B. Clerckx, Analysis and optimization of inter-tier interference coordination in downlink multi-antenna HetNets with offloading. IEEE Trans. Wirel. Commun. **14**, 6550–6564 (2015)
40. P. Henrici, *Applied and computational complex analysis* (Wiley, New York, NY, USA, 1988)
41. F. Baccelli, B. Błaszczyszyn, *Stochastic geometry and wireless networks: volume I theory* (vol. 3. Now Publishers Inc., 2009)
42. H. ElSawy, E. Hossain, M. Haenggi, Stochastic geometry for modeling, analysis, and design of multi-tier and cognitive cellular wireless networks: a survey. IEEE Commun. Surv. Tuts. **15**, 996–1019 (2013)
43. D. Commenges, M. Monsion, Fast inversion of triangular Toeplitz matrices. IEEE Trans. Autom. Control **29**(3), 250–251 (1984)
44. D. Kressner, R. Luce, Fast computation of the matrix exponential for a Toeplitz matrix, *arXiv preprint* arXiv:1607.01733 (2016)
45. M. Abramowitz, I.A. Stegun, *Handbook of mathematical functions: with formulas, graphs, and mathematical tables* (vol. 55. Courier Corporation, 1965)
46. J. Segura, Bounds for ratios of modified Bessel functions and associated Turán-type inequalities. J. Math. Anal. Appl. **374**(2), 516–528 (2011)

Chapter 4
Analysis of Multi-Antenna Wireless Networks

Abstract This chapter applies the analytical framework presented in Chap. 3 to two types of wireless networks with different propagation characteristics, namely, single-tier multi-antenna small cell networks and mm-wave wireless networks. For multi-antenna small cell networks, compared with Chap. 3, a more comprehensive analysis is performed, including the coverage probability, ASE, and energy efficiency. Closed-form expressions are derived for these metrics, which reveal the impacts of the BS density and antenna size, and also discover a phase transition phenomenon of the ASE-density relation. For mm-wave networks, both the ad hoc and cellular network models are considered, with a special focus on the unique propagation characteristics of mm-wave signals. It is shown that the coverage probabilities of both types of networks increase as a non-decreasing concave function with the antenna array size. Results presented in this chapter demonstrate the tractability and wide applicability of the general framework, as well as specific analytical methodologies for applying it to different network models.

4.1 Coverage Analysis of MISO Small Cell Networks

In Sect. 3.4.1, we have presented a general coverage analysis of multi-antenna cellular networks, where the interferer's power gain g_x is arbitrarily distributed. While general in the distribution of g_x, it loses some tractability for carrying out more comprehensive investigations, e.g., for performance metrics other than the coverage probability. In this section, we apply the general framework presented in Chap. 3 to a MISO cellular network, i.e., with multiple transmit antennas and a single receive antenna for each pair. Specifically, we shall consider a small cell scenario, where each BS has a small coverage area. Extensive systematic analysis is conducted for various performance metrics, including the coverage probability, ASE, and energy efficiency, by taking the effects of user density, BS density, and antenna size into consideration. With this more specific application, it shall be shown that the analytical framework in Chap. 3 is a powerful tool for analyzing practical wireless networks and revealing key network design insights.

X. Yu et al., *Stochastic Geometry Analysis of Multi-Antenna Wireless Networks*, https://doi.org/10.1007/978-981-13-5880-7_4

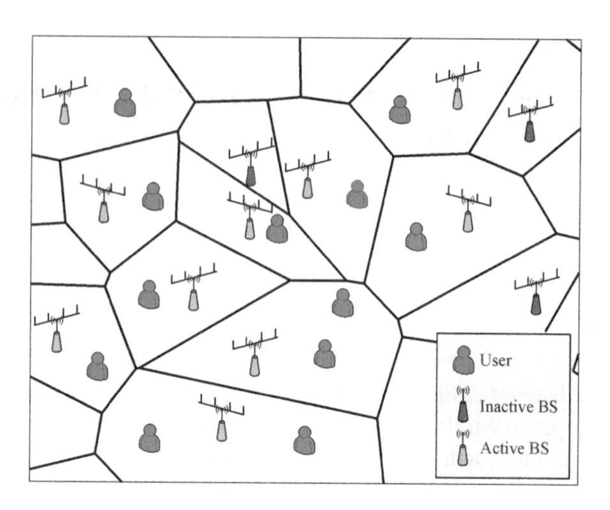

Fig. 4.1 A sample network where BSs and users are modeled as two independent PPPs. Each user connects to the closest BS

4.1.1 Network Model and Coverage Analysis

Most studies that adopt the random network model to analyze cellular networks focus on the spatial distribution of BSs, while the distribution of mobile users is largely ignored. Specifically, BSs are modeled as a PPP, and each BS always has a mobile user to serve, so the user density and the BS-user association are irrelevant, i.e., they will not affect the interference distribution and network performance. Such an assumption greatly simplifies the analysis, but it holds only when the user density is much larger than the BS density, so all the BSs will be active and have users to serve. It is generally the case for traditional sparsely deployed macrocell networks. Nevertheless, it is no longer true in small cell networks, where the user density is comparable to the BS density. Very likely there will be inactive BSs which do not have any associated user, and this factor will affect the interference distribution. In this part, we explicitly consider the user density and the BS-user association in multi-antenna cellular networks, as shown in Fig. 4.1, where BSs and users are distributed according to two independent homogeneous PPPs in $\mathbb{R}^2$, denoted as Φ_b and Φ_u, respectively. Denote the BS density as λ_b and the user density as λ_u. This system can be regarded as a single-tier network, or as one tier in a heterogeneous network with orthogonal spectrum allocation among different tiers.

We consider the downlink transmission and assume that each user is served by the nearest BS, which comprises a Voronoi tessellation relative to Φ_b. So the shape of each cell is irregular, as shown in Fig. 4.1. This kind of network model is suitable for modern cellular networks, e.g., small cell networks, where BSs are deployed irregularly [1, 2]. Due to the independent locations of BSs and users, there may be some BSs that do not have any user to serve. These BSs are called *inactive* BSs and will not transmit any signal, while BSs who have users to serve are called *active* BSs. The probability that a typical BS is active is denoted as p_a. Equivalently, p_a can be regarded as the ratio of the number of active BSs to the total number of BSs. It has been shown that p_a, as a function of the BS-user density ratio $\rho \triangleq \lambda_b/\lambda_u$, is given by [3]

$$p_a = 1 - \left(1 + \frac{1}{\mu\rho}\right)^{-\mu}, \tag{4.1}$$

where $\mu = 3.5$ [3, 4] is a constant related to the cell size distribution obtained through data fitting.[1] An active BS may have more than one user in its cell, and the BS will randomly choose one user to serve at each time slot, i.e., intracell time division multiple access (TDMA) is adopted. Note that the derivation can be easily extended to other orthogonal multiple access methods, such as FDMA [6] or SDMA [7].

We assume that each BS is equipped with M antennas,[2] while each user has a single antenna, and each BS transmits with power P_t. Universal frequency reuse is assumed, and thus each user not only receives information from its home BS but also suffers interference from all the other active BSs. Interference suppression through BS cooperation is not considered, which will be treated in Chap. 5. Moreover, due to the large scale of the small cell network, it is difficult to obtain the global channel state information (CSI) at each BS, so we assume that each active BS only has CSI of the channel to its own user.

According to (3.1), the received signal for the typical user is given by

$$\hat{s}_{x_0} = \sqrt{P_t r_0^{-\frac{\alpha}{2}}} \mathbf{h}_{x_0x_0}\mathbf{f}_{x_0}s_{x_0} + \sqrt{P_t}\sum_{x\in\tilde{\Phi}'_b} ||x||^{-\frac{\alpha}{2}}\mathbf{h}_{x_0x}\mathbf{f}_x s_x + n_{x_0}, \tag{4.2}$$

where $\mathbf{h}_{x'x} \sim \mathcal{CN}(\mathbf{0}, \mathbf{I})$ is an $1 \times M$ vector denoting the small-scale fading between the active BS located x and the user associated with the BS located at x'. In this part, we focus on MRT beamforming which has been introduced in Sect. 3.2.1, partly due to its simplicity and partly due to the fact that the optimal usage of multiple antennas in this scenario is unknown. Specified to the MISO system, the MRT beamforming vector is given by

$$\mathbf{f}_x = \frac{\mathbf{h}_{xx}}{||\mathbf{h}_{xx}||_2}. \tag{4.3}$$

In (4.2), the set of active BSs is denoted as $\tilde{\Phi}_b$. One major difficulty in the analysis is the complicated distribution of $\tilde{\Phi}_b$, which is not a simple homogeneous PPP as shown in [8], due to the coupling of the numbers of users in different cells. To simplify the following analysis, we make the same approximation as in [3], i.e., $\tilde{\Phi}_b$ is assumed to be a homogeneous PPP with density $\lambda_b p_a$. Such an approximation has been shown to be accurate in [3], and we will test its accuracy later through simulations. Note that all the following analytical results are exact for a homogeneous $\tilde{\Phi}_b$.

---

[1] Note that the value of μ can be different due to the data fitting, e.g., $\mu = 4$ was used in [5].

[2] In Chap. 3, BS antenna size was denoted by N_t and the shape parameter of the gamma distribution was denoted by M. Nevertheless, for the first application in this Chapter, we denote BS antenna size by M with a slight abuse of notation, as we shall see that these two parameters are closely related to each other.

As illustrated in Methodology 1 in Sect. 3.3.2, the first step to apply the general analytical framework is to calculate the conditional Laplace transform $\mathscr{L}(s)$. In order to derive it, we need to know the distributions of the signal power gain g_{x_0} and the interferer's power gain g_x, given in (3.16) and (3.17), which are correspondingly given by [9]

$$g_{x_0} = \left\|\mathbf{h}_{x_0 x_0}\right\|^2 \sim \text{Gamma}\left(M, 1\right), \tag{4.4}$$

and

$$g_x = \frac{\left\|\mathbf{h}_{x_0 x}\mathbf{h}_{xx}^H\right\|^2}{\left\|\mathbf{h}_{xx}\right\|^2} \sim \text{Exp}\left(1\right). \tag{4.5}$$

From (4.4), we see that, with MRT beamforming, the BS antenna size is nothing but the shape parameter of the gamma distribution of the desired signal power gain g_{x_0}. As we consider a dense network with a large number of transmitters, it is reasonable to assume an interference-limited scenario similar to that in Sect. 3.4.1, so the additive noise will be ignored in the following analysis. Later we will justify this assumption through simulations. First, we evaluate the coverage probability of the MISO small cell network, as given in the following theorem.

Theorem 4.1 *The SIR coverage probability of the MISO small cell network is given by*

$$p_c(\tau) = \left\|\left[(1 - p_a)\mathbf{I}_M + p_a\mathbf{C}_M\right]^{-1}\right\|_1, \tag{4.6}$$

with the nonzero entries in the lower triangular Toeplitz matrix $\mathbf{C}_M$ *given as*

$$c_n = \frac{\delta\tau^n}{\delta - n}\,{}_2F_1\left(n+1, n-\delta; n+1-\delta; -\tau\right), \tag{4.7}$$

for $0 \leq n \leq M - 1$.

Proof By applying Corollary 3.1 and incorporating the BS activity probability p_a in the calculation of the Laplace transform, the power series $T(z)$ in (3.94) is specified as

$$T(z) = \pi p_a \lambda_b r_0^2 \left[1 - C(z)\right]. \tag{4.8}$$

Therefore, the power series in (3.97) is written as

$$\begin{aligned}\bar{P}(z) &= \mathbb{E}_{r_0}\left[P(z)\right] = \mathbb{E}_{r_0}\left[e^{T(z)}\right] = \int_0^\infty 2\pi\lambda_b r e^{-\pi\lambda_b r^2} e^{T(z)} \mathrm{d}r \\ &= \int_0^\infty 2\pi\lambda_b r e^{-\pi\lambda[1-p_a+p_a C(z)]r^2} \mathrm{d}r = \frac{1}{1 - p_a + p_a C(z)},\end{aligned} \tag{4.9}$$

Applying Lemma 3.3 and Theorem 3.2, Theorem 4.1 can be proved.

Compared with the previous analytical results [9–12], expression (4.6) is mathematically more tractable, since we can apply the properties of the Toeplitz matrix

[13, 14] and the matrix norm for the further analysis. Denoting

$$\bar{\mathbf{P}}_M \triangleq \left[(1 - p_a)\mathbf{I}_M + p_a\mathbf{C}_M\right]^{-1}, \tag{4.10}$$

then the following lemma provides some basic properties of $\bar{\mathbf{P}}_M$ to demonstrate the tractability of the coverage probability expression.[3]

Lemma 4.1 *The matrix $\bar{\mathbf{P}}_M$ and its ℓ_1-induced norm $\|\bar{\mathbf{P}}_M\|_1$ have the following properties:*

1. *$\bar{\mathbf{P}}_M$ is a lower triangular Toeplitz matrix with positive entries, i.e.,*

$$\bar{\mathbf{P}}_M = \begin{bmatrix} \bar{p}_0 & & & & \\ \bar{p}_1 & \bar{p}_0 & & & \\ \bar{p}_2 & \bar{p}_1 & \bar{p}_0 & & \\ \vdots & & & \ddots & \\ \bar{p}_{M-1} & \cdots & \bar{p}_2 & \bar{p}_1 & \bar{p}_0 \end{bmatrix}, \tag{4.11}$$

where $\bar{p}_n > 0$, $0 \le n \le M-1$.

2. *The derivative of the norm with respect to (w.r.t.) the activity probability is given by*

$$\frac{\partial \|\bar{\mathbf{P}}_M\|_1}{\partial p_a} = \frac{1}{p_a}\left(\|\bar{\mathbf{P}}_M^2\|_1 - \|\bar{\mathbf{P}}_M\|_1\right). \tag{4.12}$$

3. *The norm $\|\bar{\mathbf{P}}_M\|_1$ is bounded as*

$$\frac{1}{1 - p_a + p_a\sum_{i=0}^{M-1}\left(1 - \frac{i}{M}\right)c_i} \le \|\bar{\mathbf{P}}_M\|_1 \le \frac{1}{1 - p_a + p_a\sum_{i=0}^{M-1}c_i}. \tag{4.13}$$

Proof See Appendix.

Note that when $M = 1$, the upper and lower bounds converge to $1/(1 - p_a + p_a c_0)$, i.e., (4.13) becomes an identity. This corresponds to the single-antenna case with Rayleigh fading, which has been shown in Remark 3.5. On the other hand, when $M \to \infty$, both bounds tend to 1.

Armed with the closed-form expression in (4.6), as well as the nice properties in Lemma 4.1, we shall carry out more refined performance analysis in the following. First of all, the third property in Lemma 4.1 directly gives an upper bound and a lower bound for the coverage probability.

Property 4.1 *The coverage probability of the MISO small cell network is bounded as*

---

[3]The reasoning of the notation $\bar{\mathbf{P}}_M$ comes from (3.95).

$$\frac{1}{1-p_{\mathrm{a}}+p_{\mathrm{a}}\sum_{i=0}^{M-1}\left(1-\frac{i}{M}\right)c_i} \leq p_{\mathrm{c}}(\tau) \leq \frac{1}{1-p_{\mathrm{a}}+p_{\mathrm{a}}\sum_{i=0}^{M-1}c_i}. \tag{4.14}$$

Proof Since $p_{\mathrm{c}}(\tau) = \left\|\bar{\mathbf{P}}_M\right\|_1$, this property follows directly from Lemma 4.1.

Note that this property separates the effect of the BS density and the effect of the BS antenna size, as p_{a} is only related to the BS-user density ratio ρ, while $\{c_i\}_{i=0}^{M-1}$ are determined by the path loss exponent and the SIR threshold. Thus, it allows us to separately analyze the impacts of the BS density and antenna size, which are presented in the following two subsections.

4.1.2 Densification: Impact of BS Density

Based on Theorem 4.1, we provide some key properties of the coverage probability for MISO small cell networks in this part. In particular, Property 4.2 provides insights on the effect of the BS density, while Property 4.3 is useful when analyzing how the BS antenna size affects system performance.

In Lemma 3.4, we have proved that the SIR invariance property still holds for cellular networks with arbitrarily distributed interferer's power gain g_x. However, it is not the case when $p_{\mathrm{a}} < 1$, i.e., when the BSs are not always active, as shown in the following result.

Property 4.2 *The coverage probability is an increase function w.r.t. the BS density, i.e.,* $\frac{\partial p_{\mathrm{c}}}{\partial \lambda_{\mathrm{b}}} \geq 0$, *and it is a constant for a given BS-user density ratio* ρ.

Proof From (4.6), we see that the coverage probability p_{c} is a function of the BS density λ_{b} through the BS activity probability p_{a}, which is a monotone decreasing function with λ_{b}. Therefore, the inequality $\frac{\partial p_{\mathrm{c}}}{\partial \lambda_{\mathrm{b}}} \geq 0$ is equivalent to $\frac{\partial p_{\mathrm{c}}}{\partial p_{\mathrm{a}}} \leq 0$. Based on Lemma 4.1, the derivative of the coverage probability with respect to p_{a} is given by

$$\frac{\partial p_{\mathrm{c}}}{\partial p_{\mathrm{a}}} = \frac{\partial \left\|\bar{\mathbf{P}}_M\right\|_1}{\partial p_{\mathrm{a}}} = \frac{1}{p_{\mathrm{a}}}\left(\left\|\bar{\mathbf{P}}_M^2\right\|_1 - \left\|\bar{\mathbf{P}}_M\right\|_1\right). \tag{4.15}$$

Since $\left\|\bar{\mathbf{P}}_M^2\right\|_1 \leq \left\|\bar{\mathbf{P}}_M\right\|_1^2$, we have

$$\begin{aligned}\frac{\partial p_{\mathrm{c}}}{\partial p_{\mathrm{a}}} &= \frac{1}{p_{\mathrm{a}}}\left(\left\|\bar{\mathbf{P}}_M^2\right\|_1 - \left\|\bar{\mathbf{P}}_M\right\|_1\right)\\ &\leq \frac{1}{p_{\mathrm{a}}}\left(\left\|\bar{\mathbf{P}}_M\right\|_1^2 - \left\|\bar{\mathbf{P}}_M\right\|_1\right)\\ &= \frac{1}{p_{\mathrm{a}}}\left\|\bar{\mathbf{P}}_M\right\|_1(1-p_{\mathrm{c}}) \leq 0,\end{aligned} \tag{4.16}$$

which is equivalent to $\frac{\partial p_{\mathrm{c}}}{\partial \lambda_{\mathrm{b}}} \geq 0$.

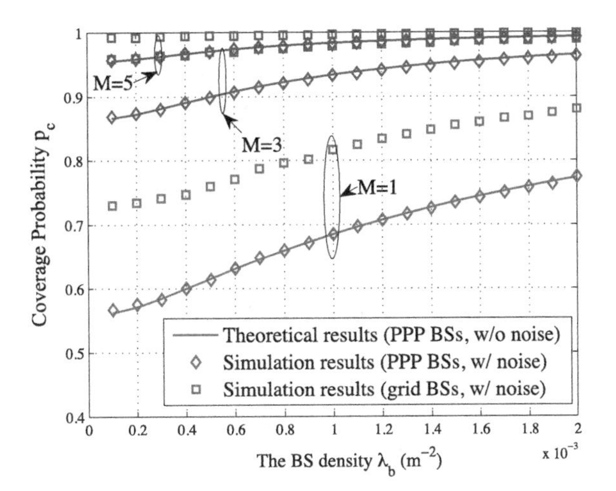

Fig. 4.2 The coverage probability versus the BS density for different numbers of BS antennas, with $\alpha = 4$, $\tau = 1$, $\lambda_u = 10^{-3}\,\text{m}^{-2}$. The transmit power is 6.3 W, and noise power considered in the simulation is $\sigma_n^2 = -97.5$ dBm. ©2014 IEEE. Reprinted, with permission, from [16]

This property implies that deploying more BSs will always increase the coverage probability for a typical user. This result actually is highly nontrivial, as increasing the BS density will increase both the signal power and the interference power. An intuitive explanation is that the average received signal power can be shown to scale with the BS density as $\lambda_b^{\frac{\alpha}{2}}$, while the average received aggregated interference power scales as $(p_a\lambda_b)^{\frac{\alpha}{2}}$. For a fixed user density, the BS activity probability p_a decreases as the BS density increases, as shown in (4.1). Therefore, the interference power increases more slowly than the signal power when the BS density increases, and thus, the coverage probability increases.

Remark 4.1 This property is the consequence of considering the explicit BS-user association and the BS activity probability. In most of the other studies, it is assumed that there is always one user for each BS to serve (i.e., $p_a = 1$), so increasing the BS density will not affect the coverage probability, i.e., the SIR invariance property holds [8]. However, in small cell networks, such as microcells and femtocells, the user density is comparable to the BS density [15], so it is necessary to take the BS activity probability into consideration. We will see more results related to the BS/user density in Sect. 4.2.

Figure 4.2 plots the coverage probability with different network models, BS densities, and antenna sizes. The analytical results without considering the noise are plotted, compared with the simulation results with noise based on the PPP model. The user density is $\lambda_u = 10^{-3}$ per square meter. We see that increasing the BS density can increase the coverage probability. It is also shown that the numerical result based on (4.6) fits the simulation results, which means that the influence of the additive noise is negligible and our approximation of $\tilde{\Phi}_b$ is accurate. To confirm our conclusions based on the random network model, we also simulate a grid-based model with the same BS density, where each cell is modeled as a hexagon. From Fig. 4.2, we find that the performance of the hexagonal cell network provides an upper bound compared to the random network model, which was also shown and explained in [8], but both network models have the same trend.

Furthermore, it can be observed that increasing the BS antenna size can also benefit the coverage probability. In particular, there is a significant gain from $M = 1$ to $M = 3$, while the gain becomes marginal from $M = 3$ to $M = 5$. In the next subsection, we shall explicitly investigate the impact of the antenna size.

4.1.3 MIMO: Impact of Antenna Size

In this part, we shall adopt the outage probability $p_{\mathrm{o}}(\tau) = 1 - p_{\mathrm{c}}(\tau)$ as the performance metric, with which more sophisticated analytical results are available. Now, we provide properties showing the effect of the antenna size, stated as follows.

Property 4.3 *The performance gain of increasing the number of BS antennas from M to $M+1$ in terms of the coverage probability is*

$$p_{\mathrm{c}}(M+1) - p_{\mathrm{c}}(M) = \bar{p}_M. \tag{4.17}$$

Denote $p_{\mathrm{o}}(M)$ as the outage probability with M antennas at each BS. We have

$$\lim_{n\to\infty} \frac{\bar{p}_n}{\bar{p}_{n+1}} = r_{\mathrm{c}}, \tag{4.18}$$

and

$$\lim_{M\to\infty} \frac{p_{\mathrm{o}}(M)}{p_{\mathrm{o}}(M+1)} = r_{\mathrm{c}}, \tag{4.19}$$

where $r_{\mathrm{c}} \in \left(1, 1+\tau^{-1}\right)$ is unrelated to M and is the solution of the equation

$${}_2F_1\left(1, -\delta; 1-\delta, (r_{\mathrm{c}}-1)\tau\right) = 1 + \frac{1}{p_{\mathrm{a}}}. \tag{4.20}$$

Proof This property can be proved in a similar way as the proof of Lemma 3.5 and Proposition 3.4.

This property has three implications:

- It is shown in (4.18) that when M is large, the benefit of adding the $(n+1)$-th antenna is $\frac{1}{r_{\mathrm{c}}}$ times smaller than adding the n-th antenna.
- Equation (4.19) implies that when M is large, the outage probability in the logarithmic scale decreases linearly with M with the slope $\log_{10}\left(\frac{1}{r_{\mathrm{c}}}\right)$, which numerically is demonstrated in Fig. 4.3. Moreover, from Fig. 4.3, we see that this linearity holds even for small values of M.

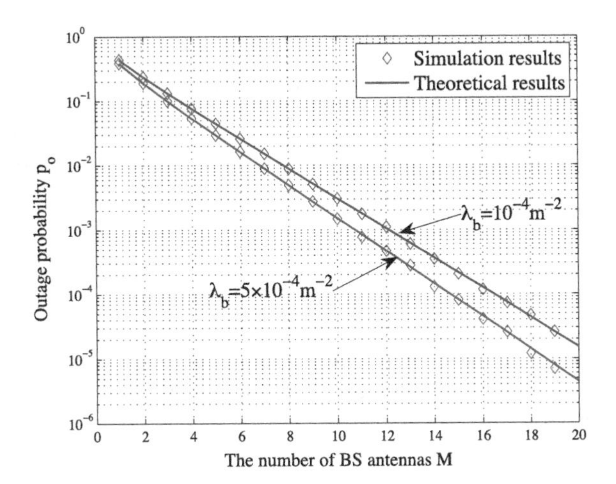

Fig. 4.3 The outage probability with different numbers of transmission antennas with $\alpha = 4$, $\tau = 1$, $\lambda_u = 10^{-3}$ per square meter. The transmit power is 6.3 W, and the noise power considered in the simulation is $\sigma_n^2 = -97.5$ dBm.

- Since increasing λ_b increases r_c, it means that the performance gain of adding one more antenna is greater with a larger BS density. As a verification, in Fig. 4.3, it is shown that the outage probability decreases faster as M increases for $\lambda_b = 5 \times 10^{-4}$ than $\lambda_b = 10^{-4}$ per square meter.

4.2 Ultra Dense Networks: ASE Versus Densification

The above discussions demonstrated the effectiveness of the analytical framework in Chap. 3 for coverage/outage probability analysis. The impact of BS density on the coverage probability with and without the consideration of the BS activity probability has been investigated in Sects. 4.1.1 and 3.5.1, respectively. On the contrary, in this section, we apply the presented analytical framework to evaluate another performance metric, i.e., network ASE, by explicitly considering the BS-user density ratio in dense multi-antenna small cell networks.

As shown in (2.26), the effect of the number of BS antennas on the ASE is the same as that on the coverage probability, which has been revealed through Property 4.3 in Sect. 4.1.3. In short, the ASE increases as we deploy more BS antennas, but the performance gain diminishes. In the following, we focus on the effect of the BS density.

Substituting (4.6) into (2.26), the network ASE is given by

$$R_a = p_a \lambda_b \left\| [(1 - p_a)\mathbf{I}_M + p_a \mathbf{C}_M]^{-1} \right\|_1 R_0. \tag{4.21}$$

Then, according to Property 4.1 of the outage probability, we get the following lower and upper bounds for the network ASE

$$\frac{\lambda_b R_0}{\frac{1}{p_a} - 1 + \sum_{i=0}^{M-1}\left(1 - \frac{i}{M}\right)c_i} \le R_a \le \frac{\lambda_b R_0}{\frac{1}{p_a} - 1 + \sum_{i=0}^{M-1} c_i}. \tag{4.22}$$

In addition to the network ASE, we also evaluate the user ASE, denoted as R_u, which is defined as the average throughput per user, given by $R_u = \rho p_a p_c R_0$. Similarly, the user ASE is bounded as

$$\frac{\rho R_0}{\frac{1}{p_a} - 1 + \sum_{i=0}^{M-1}\left(1 - \frac{i}{M}\right)c_i} \le R_u \le \frac{\rho R_0}{\frac{1}{p_a} - 1 + \sum_{i=0}^{M-1} c_i}. \tag{4.23}$$

Note that these bounds are irrelevant to the BS activity probability p_a. In the following, we will investigate the impact of the BS density on the network and user ASEs in two different scenarios. We will see that these two throughput metrics are affected by the BS/user density in different ways, and subsequently, important design guidelines will be drawn.

Scenario 1 (For a fixed user density)

We first consider a fixed user density and investigate how the network ASE changes with the BS density. This is of practical relevance, as it corresponds to investigating how much additional gain can be provided if the operator deploys more BSs. In this case, the effect of the BS density on the network ASE R_a is the same as that on the user ASE R_u. We consider the following three different regimes in terms of the BS density.

- **Low BS density regime**: $\lambda_b \ll \lambda_u$ or $\rho \ll 1$, so $p_a \approx 1$ (e.g., $p_a > 0.99$ when $\rho \le 0.1$), which means almost all the BSs are active. From (4.21), we see that in this regime R_a increases *linearly* with λ_b, i.e.,

$$R_a = \tilde{c}_0 \lambda_b, \tag{4.24}$$

  where $\tilde{c} \triangleq \left\| [(1 - p_a)\mathbf{I}_M + p_a \mathbf{C}_M]^{-1} \right\|_1 R_0$ is unrelated to λ_b. This is the common case considered in most previous works, such as [8].
- **High BS density regime**: $\lambda_b \gg \lambda_u$, so all the users are being served and $p_a \approx \frac{1}{\rho}$. From (4.22), we see that increasing the BS density can increase the ASE, but the improvement is quite limited. In particular, the upper and lower bounds in (4.22) are given by

$$\frac{\lambda_u R_0}{1 + \frac{\sum_{i=0}^{M-1}(1-\frac{i}{M})c_i - 1}{\rho}} \le R_a \le \frac{\lambda_u R_0}{1 + \frac{\sum_{i=0}^{M-1} c_i - 1}{\rho}}. \tag{4.25}$$

  Therefore,

$$R_a \to \lambda_u R_0 \quad \text{when} \quad \rho \gg 1. \tag{4.26}$$

  So there is no need to further increase λ_b. This special case is considered in [3].
- **Medium BS density regime**: $\lambda_b \sim \lambda_u$, i.e., the BS density and the user density are comparable, which is a more practical case for small cell networks [15]. For this

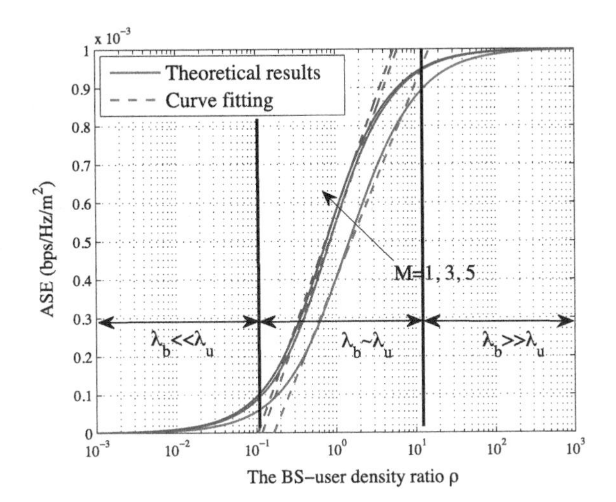

Fig. 4.4 The network ASE for different BS-user density ratios, with $\alpha = 4$ and $\tau = 1$. ©2014 IEEE. Reprinted, with permission, from [16]

case, an exact expression for the network ASE is given in (4.21), but it is difficult to get the scaling result in this finite regime. We therefore resort to data fitting, which shows that R_{a} increases *logarithmically* with the BS density, i.e.,

$$R_{\mathrm{a}} \approx \tilde{c}_1 \log \rho + \tilde{c}_2, \tag{4.27}$$

where $\tilde{c}_1$ and $\tilde{c}_2$ can be determined by data fitting. Some numerical examples are shown in Fig. 4.4 to validate such relationship. It will be interesting to analytically characterize this observed behavior.

The above analysis shows that the BS-user density ratio is critical to network ASE. Thus, it should be carefully taken into consideration when evaluating dense multi-antenna networks.

Scenario 2 (With a fixed ρ)

In this case, the BS density varies in proportion to the user density. From (4.23), we see that the user ASE R_{u} is the same for a fixed ρ; while from (4.22), we see that R_{a} increases linearly with λ_{b} for a fixed ρ. This means that if we keep the BS-user density ratio fixed, the network ASE grows linearly with the BS density, while the ASE of a typical user stays the same. Equivalently, this indicates that once the user density increases, the operator can improve the network ASE by deploying more small BSs, while maintaining the QoS for each user, which demonstrates the advantage of small cell networks.

4.3 Green Networking: Energy Efficiency Analysis

In addition to the three evolution paths introduced in Sect. 1.2, green communications is drawing more and more attention on a global scale. Achieving higher energy efficiency is among the main design objectives of the next-generation cellular networks. In [17], it was pointed out that BSs consume more than 60% of the total energy in cellular networks. As more and more BSs are deployed, the effect of network densification on the energy efficiency should be carefully quantified. In the following, we analyze the network energy efficiency in multi-antenna small cell networks, which will be shown to depend critically on the BS power consumption model. In particular, there is no simple monotonic result with respect to the BS density or multiple BS antennas, and different conclusions will be drawn under different conditions.

4.3.1 *Densification: Impact of BS Density*

The energy efficiency metric, as well as the BS power consumption model, has been introduced in Sect. 2.2. By substituting (4.6) into (2.30), we obtain the following expression of the energy efficiency:

$$\xi_{\mathrm{EE}} = \frac{\left\| \left[\left(\frac{1}{p_{\mathrm{a}}} - 1 \right) \mathbf{I}_M + \mathbf{C}_M \right]^{-1} \right\|_1 R_0}{p_{\mathrm{a}} \left(\frac{1}{\xi} P_{\mathrm{t}} + M P_{\mathrm{c}} \right) + P_0}. \tag{4.28}$$

Then we get the following result showing the effect of the BS density.

Proposition 4.1 *The energy efficiency is a decreasing function with* λ_{b} *if*

$$\frac{P_0}{P_{\mathrm{BS}}} \geq 1 - \frac{\left\| \mathbf{C}_M^{-2} \right\|_1}{\left\| \mathbf{C}_M^{-1} \right\|_1} \triangleq \tau_{P_0}, \tag{4.29}$$

where $P_{\mathrm{BS}} = \frac{1}{\eta} P_{\mathrm{t}} + M P_{\mathrm{c}} + P_0$. *Otherwise, the energy efficiency first increases and then decreases as* λ_{b} *increases, and there is a nonzero optimal BS density* $\lambda_{\mathrm{b}}^\star$ *that maximizes the energy efficiency. The approximated maximum energy efficiency is*

$$\xi_{\mathrm{EE}}^\star \approx \frac{R_0}{\left(\sqrt{\frac{1}{\xi} P_{\mathrm{t}} + M P_{\mathrm{c}}} + \sqrt{P_0 B} \right)}, \tag{4.30}$$

where

$$\sum_{i=0}^{M-1} c_i - 1 \leq B \leq \sum_{i=0}^{M-1} \left(1 - \frac{i}{M} \right) c_i - 1, \tag{4.31}$$

and the corresponding optimal BS density is

$$\lambda_b^\star \approx \frac{1}{\mu}\left[1-\left(1-\sqrt{\frac{P_0}{B\left(\frac{1}{\xi}P_t+MP_c\right)}}\right)^{-\frac{1}{\mu}}\right]\lambda_u. \tag{4.32}$$

Proof Since p_a is a monotone decreasing function with λ_b, the effect of the BS density λ_b on the energy efficiency is the opposite as that of the BS activity probability p_a. Therefore, the condition (4.29) is derived by investigating the derivative of (4.28) w.r.t. the BS activity probability p_a, while the approximated optimal energy efficiency and the corresponding BS density can be obtained through (4.14).

From this result, we see that the non-transmission power consumption P_0 plays a critical role in the energy efficiency. Particularly, when $P_0/P_{BS} \geq \tau_{P_0}$, increasing the BS density always decreases energy efficiency, although it can improve the ASE. On the other hand, when $P_0/P_{BS} < \tau_{P_0}$, there is a nonzero BS density that can achieve the maximum energy efficiency, which is instructive when designing and operating a cellular network. Typical values of the power for different types of BSs can be found in [18], e.g., the typical transmit power for micro-BS, pico-BS, and femto-BS are 6.3 W, 0.13 W, and 0.05 W, respectively.

4.3.2 *MIMO: Impact of Antenna Size*

We have seen that increasing the BS antenna size increases the ASE, but it also consumes more circuit power P_c. Next, we investigate how the BS antenna number affects the overall network energy efficiency. Denote $\xi_{EE}(M)$ as the energy efficiency with M antennas per BS, the energy efficiency is written as

$$\xi_{EE}(M) = \frac{\sum_{n=0}^{M-1}\bar{p}_n}{\frac{1}{\xi}P_t + MP_c + \frac{P_0}{p_a}}R_0. \tag{4.33}$$

Then the effect of the number of BS antennas on the energy efficiency is given in the following proposition.

Proposition 4.2 *There is an optimal number of BS transmit antennas $M^\star$ that maximizes the energy efficiency. When $M > M^\star$, increasing M will decrease the energy efficiency, while for $M < M^\star$, deploying more antennas can improve the energy efficiency. The optimal $M^\star$ is the greatest integer that is smaller than the solution of the equation*

$$F(M) = \frac{p_a\left(\frac{1}{\xi}P_t\right)+P_0}{p_aP_c}, \tag{4.34}$$

where $F(M) \triangleq \frac{p_c(M)}{\bar{p}_{M-1}} - M$.

Proof See Appendix.

Since $F(M) = \frac{1}{\bar{p}_{M-1}} \sum_{n=0}^{M-2} (\bar{p}_n - \bar{p}_{M-1})$, it is obvious that $F(M)$ is an increasing function with M. Thus, if we could deploy BSs with a smaller P_c, the optimal antenna size would be larger. Subsequently, both the spectral efficiency and the energy efficiency can be improved.

An extreme case is $M^\star = 1$, which implies that deploying single-antenna BSs provides higher energy efficiency than using multi-antenna BSs. For this case, we find the condition from (4.75) as $P_c \geq \frac{k_1\left(p_a \frac{1}{\xi} P_t + P_0\right)}{1+(k_0-k_1)p_a}$, where the right-hand side of this inequality is a monotone function w.r.t. p_a, which means if the condition

$$P_c \geq \max\left(k_1 P_0, \frac{k_1\left(\frac{1}{\xi}P_t + P_0\right)}{1+k_0-k_1}\right) \triangleq \tau_{P_c} \tag{4.35}$$

is satisfied, for any BS and user densities, deploying single-antenna BSs is more energy efficient than multi-antenna BSs. Therefore, multi-antenna BSs are preferable in terms of energy efficiency only when the circuit power consumption is smaller than the threshold τ_{P_c}. To summarize, Table 4.1 shows the main results on the effects of the BS density and BS antenna size on the network ASE and energy efficiency.

To illustrate the analytical results, Fig. 4.5 shows the change of the network energy efficiency as the BS density increases. The system setting is the same as Fig. 4.2. For the power consumption model, we consider a micro-BS with $\xi = 0.32$, $P_t = 6.3$ W, $P_c = 35$ W, and $P_0 = 34$ W [19]. By substituting these values, we find that the condition (4.35) is satisfied, which means it is more energy efficient to deploy single-antenna BSs. Moreover, for $M = 1$, there is a nonzero optimal BS density, which can be calculated from (4.32) as $\lambda_b^\star = 0.32\lambda_u$. On the other hand, when $M > 2$, the

Table 4.1 The effects of λ_b and M on R_a and ξ_{EE}. © 2014 IEEE. Reprinted, with permission, from [16]

	Network ASE	Energy efficiency
λ_b	for fixed λ_u: $\begin{cases} R_a = c_0\lambda_b & \text{for } \lambda_b \ll \lambda_u \\ R_a \approx c_1 \log\lambda_b + c_2 & \text{for } \lambda_b \sim \lambda_u \\ R_a \to \lambda_u R_0 & \text{for } \lambda_b \gg \lambda_u \end{cases}$ for fixed ρ: R_u is fixed, while R_a is linear w.r.t. λ_b	• If $\frac{P_0}{P_{BS}} \geq \tau_{P_0}$, ξ_{EE} decreases with λ_b • Otherwise, there is one $\lambda_b^\star$ to maximize ξ_{EE}
M	$R_a = \lambda_b p_a R_0 \sum_{n=0}^{M-1} \bar{p}_n$, where $\lim_{n\to\infty} \frac{\bar{p}_n}{\bar{p}_{n+1}} = r_u$	• If $P_c \geq \tau_{P_c}$, single-antenna BSs achieve the maximum ξ_{EE} • Otherwise, there is one $M^\star > 1$ to maximize ξ_{EE}

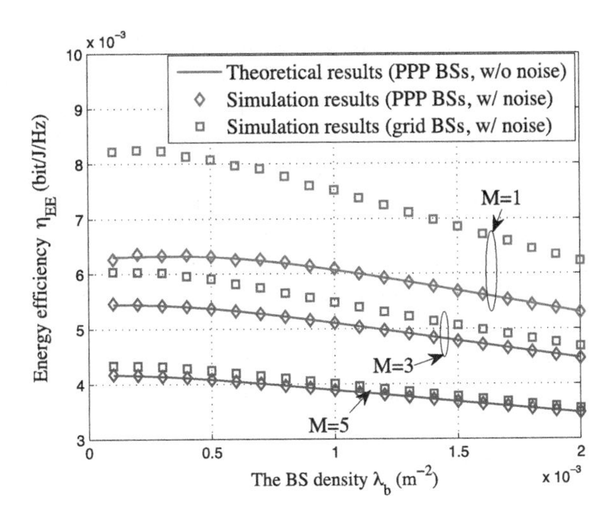

Fig. 4.5 Energy efficiency versus BS density for different numbers of BS antenna with $\alpha = 4$, $\tau = 1$, $\lambda_u = 10^{-3}\text{m}^{-2}$, $\xi = 0.32$, $P_t = 6.3$ W, $P_c = 35$ W, $P_0 = 34$ W, and the noise power considered in simulation is $\sigma_n^2 = -97.5$ dBm. In the figure, it is shown that the optimal BS density for $M = 1$ case is about $0.3 \times 10^{-3}\text{m}^{-2}$ for both the random and grid-based models. ©2014 IEEE. Reprinted, with permission, from [16]

energy efficiency is a decreasing function w.r.t. λ_b. These results are confirmed by the simulation results in Fig. 4.5. In addition, we find that compared with the random network model, the performance of the hexagonal cell network still provides an upper bound, while the trends of both network models are the same. Even the optimal BS density $\lambda_b^\star$ for a grid-based model is close to our analytical result. Interestingly, the analytical result for the random network model gets closer to the grid-based model as M increases.

4.4 New Spectrum: Millimeter-Wave Networks

As introduced in Sect. 1.2.3, uplifting the carrier frequency to mm-wave bands has been proposed as one promising way to boost the network capacity. Nevertheless, the design and deployment of mm-wave communication systems face a few key challenges. Lately, channel measurements have confirmed some unique propagation characteristics of mm-wave signals [20], which shall significantly affect the performance of mm-wave networks. In particular, mm-wave signals are sensitive to blockages, which causes totally different path loss laws for line-of-sight (LOS) and non-line-of-sight (NLOS) mm-wave signals. Furthermore, diffraction and scattering effects have been shown to be limited for mm-wave signals. This makes the conventional channel model for sub-6 GHz systems no longer suitable, and thus, more sophisticated channel models are needed for the performance analysis of mm-wave networks.

Another distinguishing characteristic of mm-wave signals is the directional transmission. Thanks to the small wavelength of mm-wave signals, large-scale directional antenna arrays can be leveraged to provide substantial array gains and synthesize highly directional beams, which help to compensate for the additional free space path loss caused by the tenfold increase of the carrier frequency [21]. More importantly, different from the rich diffraction and scattering environment in sub-6 GHz systems, directional antennas will dramatically change the signal power, as well as the interference power. In mm-wave networks, the signal or interference power is highly directional and closely related to the angles of departure/arrival (AoDs/AoAs). In particular, the directional antenna array will provide variable power gains corresponding to different AoDs/AoAs. Even a slight shift of AoD/AoA may lead to a large array gain variation. Therefore, it is necessary and critical to incorporate the directional antenna arrays when analyzing mm-wave networks. In this section, we shall investigate the performance of mm-wave wireless networks, taking account of both mm-wave signal propagation characteristics and directional antenna arrays. While it appears that the mm-wave network is quite different from the general network model analyzed in Chap. 3, we shall show that the general analytical framework is still applicable.

4.4.1 Characteristics of Mm-Wave Networks

We consider downlink transmission. Both the mm-wave ad hoc and cellular networks are investigated. We first present the common features for both types of networks, and the difference will be specified later. The transmitters are assumed to be distributed according to a homogeneous PPP [22]. As depicted in Fig. 4.6a, we consider the receiver at the origin, which, under an expectation over the point process, becomes the typical receiver. We assume that each receiver has a single receive antenna and is receiving signals from the corresponding transmitter equipped with a directional antenna array composed of N_t elements.

LOS and NLOS Signals

We use the LOS ball [24, 25] to model the blockage effect as shown in Fig. 4.6a. Specifically, we define an LOS radius R, which represents the distance between a receiver and its nearby blockages, and the LOS probability of a certain link is one within R and zero outside the radius. Compared with other blockage models adopted in the performance analysis for mm-wave networks, e.g., the 3GPP-like urban microcellular model, the LOS ball model has a better fit with real-world blockage scenarios [26]. The incorporation of the blockages induces different path loss laws for LOS and NLOS links. It has been pointed out in [24, 27] that NLOS signals and NLOS interference are negligible in mm-wave networks. Hence, we focus on the analysis where the typical receiver is associated with an LOS transmitter and the interference stems from LOS interferers. The relevant transmitters thus form a PPP, denoted as Φ, with density λ_b in a disk of radius R centered at the origin. Next, we will justify the LOS assumption through simulations.[4]

---

[4]Without specific indication, the default simulation settings for mm-wave networks are as follows. We assume that the bandwidth is 1 GHz, and the transmit power of each BS is set as 1 W. The

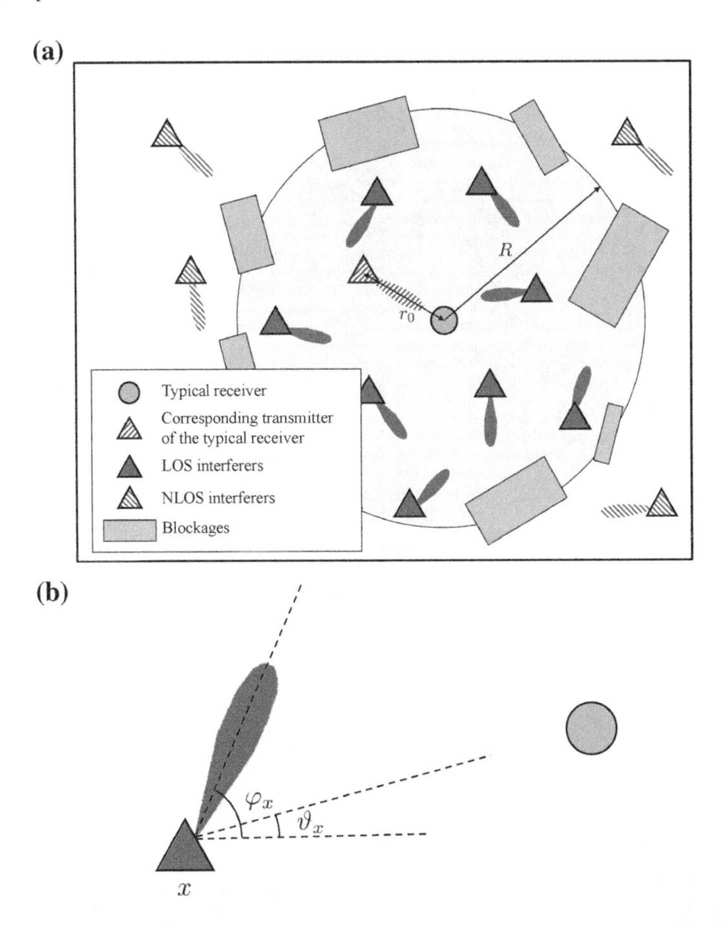

Fig. 4.6 **a** A sample mm-wave network where transmitters are modeled as a PPP. The LOS ball is used to model the blockage effect in the network. **b** Illustration of the spatial AoDs ϑ_x and φ_x. ©2017 IEEE. Reprinted, with permission, from [23]

In Fig. 4.7a, we show a simulation of the SINR coverage probability without incorporating the NLOS serving BS and NLOS interferers, whose curve almost coincides with that including NLOS components. This demonstrates that the impact of NLOS signals and interference is negligible and validates the LOS assumption, i.e., we only need to focus on the analysis where the typical receiver is associated with an LOS transmitter and the interference is brought by LOS interferers. The underlying reasons are as follows for different BS densities:

---

separation between the antenna elements is $d = \frac{\lambda}{4}$, i.e., quarter-wavelength to avoid the grating lobes. From the recent measurements of mm-wave signal propagations [20], the path loss exponent α is close to 2 and the intercept is $\beta = -61.4$ dB. All the simulation results shown in this section are averaged over 5×10^5 realizations.

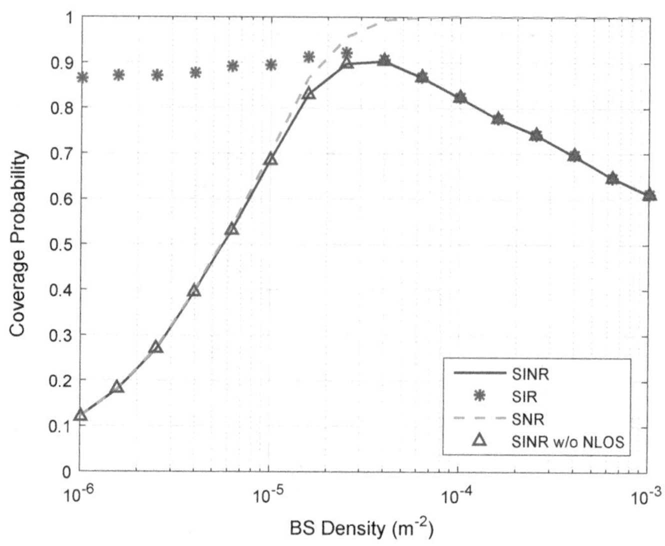

(a) SINR, SIR, and SNR coverage probabilities in mm-wave cellular networks when $R = 200$ m, $N_t = 64$, $\tau = 10$ dB, $M = 3$, and $\alpha = 2.1$.

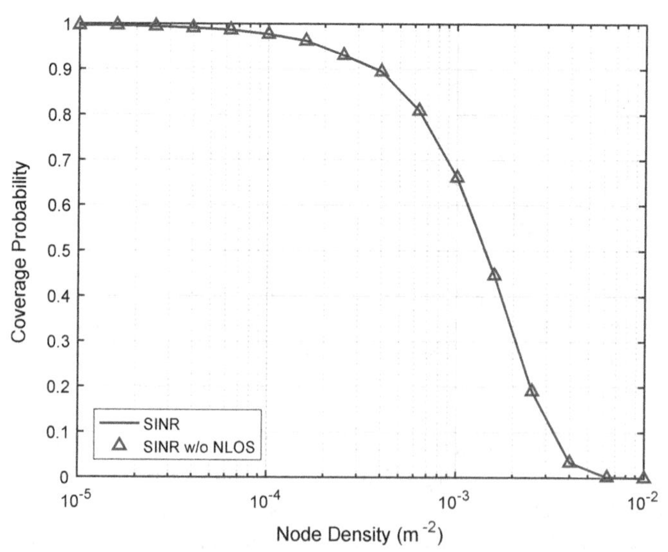

(b) SINR, SIR, and SNR coverage probabilities in mm-wave ad hoc networks when $R = 180$ m, $N_t = 64$, $\tau = 5$ dB, $M = 5$, $\alpha = 2.2$, and $r_0 = 25$ m.

Fig. 4.7 The impact of NLOS signals and interference in mm-wave **a** cellular networks and **b** ad hoc networks. ©2017 IEEE. Reprinted, with permission, from [23]

- When the BS density is low, the network is operating in the noise-limited regime, and thus, only the LOS signal matters.
- At medium BS densities, there is a certain probability to have an LOS serving BS, and the interference gradually affects the SINR coverage. However, the LOS interference power is much higher compared to the NLOS ones. On the other hand, when the typical link is NLOS, it is difficult to achieve a satisfactory SINR value
- Very dense mm-wave networks will be LOS interference-limited, which has been investigated in [24, 28].

Figure 4.7b demonstrates the impact of NLOS interferers when the tagged transmitter is in the LOS condition. It manifests that, with an LOS transmitter associated with the typical receiver, the NLOS interference is also negligible for the reasons which are similar to those in cellular networks [27]. Hence, it is reasonable to neglect the NLOS components in the analysis of mm-wave networks. Although we showed that NLOS parts are minor, note that all the simulations include them to maintain the completeness and consistency.

In addition, the effects of noise and interference in mm-wave networks are also investigated. In Fig. 4.7a, we evaluate the SIR and SNR coverage probabilities versus the BS density in mm-wave cellular networks. It was found in [25] that the SIR coverage probability will monotonically decrease with the increasing BS density in sub-6 GHz networks with the dual-slope path loss model, which, however, no longer holds in mm-wave cellular networks. This is because the difference in the small-scale fading for LOS and NLOS propagations in mm-wave networks, which were assumed to be the same in [25]. When the BS density gradually increases, the signal link tends to experience Nakagami fading rather than Rayleigh fading. This change in small-scale fading results in a slight increase of the SIR coverage probability, which also implicitly illustrates that Nakagami fading provides better coverage than Rayleigh fading. Therefore, as a lower bound of both SIR and SNR coverage probabilities, the SINR coverage probability in mm-wave cellular networks has a peak value with the increasing BS density. On the other hand, different from mm-wave cellular networks, the SINR coverage probability decreases with network densification due to the fixed dipole distance and arbitrarily close interferers, which is shown in Fig. 4.7b. This evaluation indicates the importance of analyzing the SINR distribution in mm-wave cellular networks, while the SIR coverage can be used as a good metric for mm-wave ad hoc networks.

Propagation Model and Analog Beamforming

Directional antenna arrays are leveraged to provide significant beamforming gains to overcome the path loss and to synthesize highly directional beams. Universal frequency reuse is assumed, and thus, according to (3.1), the received signal for the typical receiver is given by

$$\hat{s}_{x_0} = \sqrt{\beta} r_0^{-\frac{\alpha}{2}} \mathbf{h}_{x_0} \mathbf{w}_{x_0} \sqrt{P_t} s_{x_0} + \sum_{x \in \Phi'} \sqrt{\beta} \|x\|^{-\frac{\alpha}{2}} \mathbf{h}_x \mathbf{w}_x \sqrt{P_t} s_x + n_{x_0}, \tag{4.36}$$

where the path loss exponent and intercept are symbolized by α and β [20], and the channel vector between the interferer and the typical receiver is denoted as $\mathbf{h}_x \in \mathbb{C}^{1\times N_t}$. Due to high free space path loss, the mm-wave propagation environment is well characterized by a clustered channel model, i.e., the Saleh–Valenzuela model [20],

$$\mathbf{h}_x = \sqrt{N_t}\sum_{l=1}^{L} \omega_{xl}\mathbf{a}_t^H(\vartheta_{xl}), \tag{4.37}$$

where L is the number of clusters. The complex small-scale fading gain of the l-th cluster is denoted as ω_{xl}. Due to the poor scattering environment, especially for LOS signals and interference, the Rayleigh fading assumption commonly used in sub-6 GHz systems no longer holds, which has also been noted in recent works [26]. We assume, as in [24], that $|\omega_{xl}|$ follows independent Nakagami-M fading for each link.[5]

For mm-wave channels containing LOS components, the effect of NLOS signals is negligible since the channel gains of NLOS paths are typically 20 dB weaker than those of LOS signals [20]. Hence, for the remainder of this chapter, we will focus on LOS paths, i.e., $L = 1$, and adopt a uniformly random single path (UR-SP) channel model that is commonly used in mm-wave network analysis [29–32]. In addition, $\mathbf{a}_t(\vartheta_x)$ represents the transmit array response vector corresponding to the spatial AoD ϑ_x. We consider the uniform linear array (ULA) with N_t antenna elements. Therefore, the array response vectors are written as

$$\mathbf{a}_t(\vartheta_x) = \frac{1}{\sqrt{N_t}}\left[1, \ldots, e^{j2\pi k\vartheta_x}, \ldots, e^{j2\pi(N_t-1)\vartheta_x}\right]^T, \tag{4.38}$$

where $\vartheta_x = \frac{d}{\lambda}\cos\phi_x$ is assumed uniformly distributed over $\left[-\frac{d}{\lambda}, \frac{d}{\lambda}\right]$, and $0 \le k < N_t$ is the antenna index. Furthermore, d, λ, and ϕ_x are the antenna spacing, wavelength, and physical AoD. In order to enhance the directionality of the beam, the antenna spacing d should be no larger than the half-wavelength to avoid grating lobes [33].

While various space-time processing techniques can be applied at each multi-antenna mm-wave transmitter, we focus on analog beamforming, where the beam direction is controlled via phase shifters. Due to the low cost and low power consumption, analog beamforming has already been adopted in commercial mm-wave systems, such as WiGig (802.11ad) [21]. Assuming the spatial AoD of the channel between the transmitter at location x and its serving user is φ_x, the optimal analog beamforming vector is well known and given by

$$\mathbf{w}_x = \mathbf{a}_t(\varphi_x), \tag{4.39}$$

[5] In previous sections, the notation M was used to represent the shape parameter of the gamma distribution. With Nakagami-M fading, the signal power gain is gamma distributed as $g_{x_0} \sim \text{Gamma}(M, 1/M)$.

which means the transmitter should align the beam direction exactly with the AoD of the channel to obtain the maximum power gain.

As shown in Fig. 4.6b, based on the optimal analog beamforming vector (4.39), for the typical receiver, the product of small-scale fading gain and beamforming gain of the transmitter at location x is given by

$$|\mathbf{h}_x \mathbf{w}_x|^2 = N_\mathrm{t} \, |\omega_x|^2 \left|\mathbf{a}_\mathrm{t}^H(\vartheta_x)\mathbf{a}_\mathrm{t}(\varphi_x)\right|^2, \tag{4.40}$$

where $|\omega_x|^2$ is the power gain of small-scale fading. By defining the array gain function $G_{\mathrm{act}}(x)$ as

$$G_{\mathrm{act}}(x) \triangleq \frac{\sin^2(\pi N_\mathrm{t} x)}{N_\mathrm{t}^2 \sin^2(\pi x)}, \tag{4.41}$$

the normalized array gain of the transmitter at location x can be expressed as

$$\begin{aligned} \left|\mathbf{a}_\mathrm{t}^H(\vartheta_x)\mathbf{a}_\mathrm{t}(\varphi_x)\right|^2 &= \frac{1}{N_\mathrm{t}^2}\left|\sum_{i=0}^{N_\mathrm{t}-1} e^{j2\pi i(\vartheta_x - \varphi_x)}\right|^2 \\ &= \frac{\sin^2[\pi N_\mathrm{t}(\vartheta_x - \varphi_x)]}{N_\mathrm{t}^2 \sin^2[\pi(\vartheta_x - \varphi_x)]} = G_{\mathrm{act}}(\vartheta_x - \varphi_x), \end{aligned} \tag{4.42}$$

where ϑ_x and φ_x are independent uniformly distributed random variables over $\left[-\frac{d}{\lambda}, \frac{d}{\lambda}\right]$. The array gain function in (4.42) is a normalized *Fejér kernel* with factor $\frac{1}{N_\mathrm{t}}$ and is referred as the *actual antenna pattern*. In fact, the distribution of $\vartheta_x - \varphi_x$ in (4.42) is uniform. Note that this substitution will not change the overall distribution of the array gain, as verified in the following lemma.

Lemma 4.2 *The array gain $G_{\mathrm{act}}(\vartheta_x - \varphi_x)$ is equal in distribution to $G_{\mathrm{act}}\left(\frac{d}{\lambda}\theta_x\right)$, where θ_x is a uniformly distributed random variable over* $[-1, 1]$.

Proof The proof is based on the uniform distribution of ϑ_x and φ_x, and the periodic property of the function $e^{j2\pi x}$ in (4.42). The proof has been established in [29, Appendix A].

Although the Fejér kernel has a relatively simple analytical form, it does not lend itself to further analysis due to the sine functions in both the numerator and denominator, which calls for an approximate antenna pattern with both accuracy and tractability in performance analysis of mm-wave networks. Next, we introduce two new approximate antenna patterns, as well as the flat-top antenna pattern, which has been widely used in existing works. Figure 4.8 visualizes these antenna patterns and evaluates the coverage probabilities with different antenna patterns through simulation.

(1) *Flat-top antenna pattern*: Most of the studies on mm-wave networks [24, 27, 34, 35] adopt this simplified antenna pattern in the coverage analysis, where the array gains within the half-power beamwidth (HPBW) [33] are assumed to be the

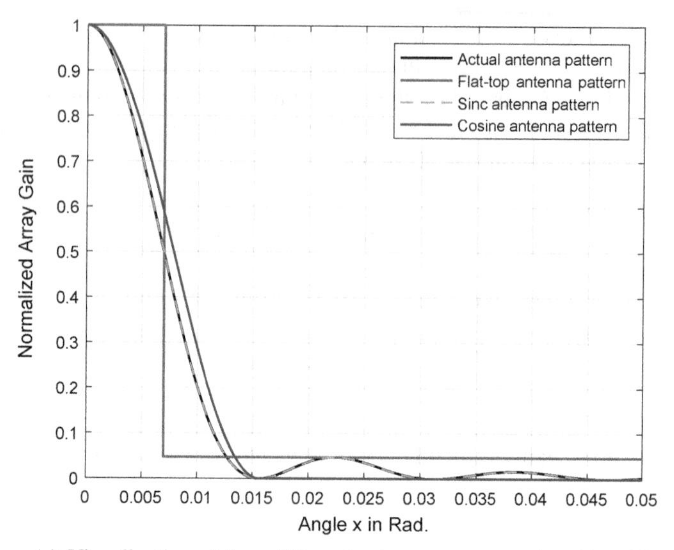

(a) Visualization of four different antenna patterns when $N_t = 64$.

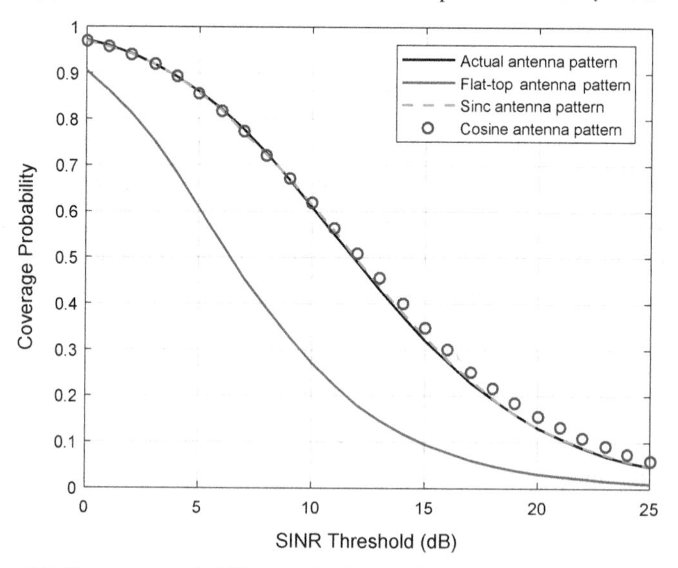

(b) Coverage probability evaluations using four different antenna patterns in mm-wave cellular networks when $R = 200$ m, $N_t = 64$, $\lambda_b = 1 \times 10^{-3}$ m^{-2}, $M = 3$, and $\alpha = 2.1$.

Fig. 4.8 The comparisons between different approximate antenna patterns. ©2017 IEEE. Reprinted, with permission, from [23]

maximum power gain, and the array gains corresponding to the remaining AoDs are approximated to be the first minor maximum gain of the actual antenna pattern. While this simple approximation is highly tractable, it introduces huge discrepancies when we evaluate the network coverage probability, as shown in Fig. 4.8b.[6]

(2) *Sinc antenna pattern*: Instead of the actual antenna pattern, a tight lower bound is widely adopted for the numerical analysis in antenna theory. Since the antenna spacing d is usually no larger than half-wavelength to avoid grating lobes, and $\sin x \simeq x$ for small x, the array gain function can be approximately expressed as [33, Eq. (6–10d)]

$$G_{\text{sinc}}(x) \triangleq \frac{\sin^2(\pi N_t x)}{(\pi N_t x)^2}, \tag{4.43}$$

which is a squared sinc function. The accuracy of this tight lower bound is shown in [33, Appendix I, II]. In Fig. 4.8a, it shows that the sinc antenna pattern is almost the same as the actual antenna pattern, and there is almost no error when using this approximate antenna pattern to investigate the coverage probability, as illustrated in Fig. 4.8a. Moreover, note that the sinc function is more tractable due to the absence of the sine function in the denominator, compared to the actual antenna pattern.

(3) *Cosine antenna pattern*: Another antenna pattern approximation is based on the cosine function as follows:

$$G_{\cos}(x) = \begin{cases} \cos^2\left(\frac{\pi N_t}{2} x\right) & |x| \le \frac{1}{N_t}, \\ 0 & \text{otherwise}, \end{cases} \tag{4.44}$$

where the nonzero part is an elementary function with better analytical tractability. In Fig. 4.8a, we observe that the cosine antenna pattern provides a good approximation for the main lobe gains while sacrificing the accuracy for the side lobe ones. When incorporated into the coverage probability, the cosine antenna pattern has a negligible gap between the actual antenna pattern, which can be viewed as a desirable trade-off between accuracy and tractability in performance analysis for mm-wave networks.

Figure 4.8 shows that the sinc and cosine antenna patterns are more accurate than the flat-top pattern. In particular, they are superior to the flat-top pattern since they accurately capture the impact of directional antenna arrays in mm-wave networks. In particular, given the operating frequency and the antenna spacing, the antenna pattern is critically determined by the array size. In the flat-top pattern, however, it is very difficult to quantitatively and accurately depict the variation of the HPBW and the first minor maximum for different array sizes and AoDs. Moreover, the binary quantization of the array gains cannot reflect the roll-off characteristic of the actual antenna pattern and therefore is unable to provide different array gains for various AoDs. In other words, the flat-top antenna pattern obliterates the possibility of analyzing the impact of directional antenna arrays, which is a critical and unique

[6]The gap can be narrowed by heuristically choosing different parameters for the flat-top pattern, e.g., beamwidth and front-back ratio, but the overall shape of the coverage probability remains different.

issue in mm-wave systems. On the contrary, the sinc and cosine antenna patterns are explicit functions of the array size, which makes it possible to investigate the relation between the coverage probability and the directional antenna arrays. The sinc and cosine antenna patterns will be adopted in the coverage analysis for mm-wave ad hoc and cellular networks in the following section.

4.4.2 Coverage Analysis of Mm-Wave Networks

In the previous sections in this chapter, the analytical framework in Chap. 3 has been applied to dense multi-antenna networks at sub-6 GHz bands. With practical MRT beamforming, the interferer's power gain g_x is gamma distributed. However, with analog beamforming in mm-wave networks, the distribution becomes more complicated due to the directional antenna gain incorporated in the interferer's power gain. In the following, we shall show that the analytical framework can be applied to analyze the more complex case in mm-wave networks, and, more importantly, key network insights are also unraveled. Similar to Sect. 2.4, we consider two network models for mm-wave networks, i.e., mm-wave ad hoc and cellular networks.

Mm-Wave Ad Hoc Networks

Millimeter-wave communications has been proposed as a promising technique for next-generation ad hoc networks with short-range transmission, e.g., military battlefield networks [36], high-fidelity video transmission [21], and D2D networks [37]. In this section, we first derive an analytical expression of the coverage probability for mm-wave ad hoc networks, based on which we then investigate the critical role of directional antenna arrays in such networks.

In mm-wave ad hoc networks, we assume that the typical dipole pair is in the LOS condition, i.e., $r_0 \leq R$. In fact, if the typical receiver is associated with an NLOS transmitter out of the LOS radius, due to the huge path loss and high noise power at mm-wave bands, the coverage probability will be fairly low (close to zero) for a practical SINR threshold, and therefore with little analytical significance. Furthermore, in ad hoc networks, the nearest interferer can be arbitrarily close to the typical receiver, and the received SINR is expressed by (3.18), where $\sigma_{\mathrm{n}}^2 = \frac{\sigma^2}{\beta P_{\mathrm{t}} N_{\mathrm{t}}}$ is the normalized noise power. The signal power gain g_{x_0} and the interferer's power gain g_x, given in (3.16) and (3.17), which are correspondingly determined as

$$g_{x_0} = |\omega_{x_0}|^2, \tag{4.45}$$

and

$$g_x = |\omega_x|^2 G_{\mathrm{act}}\left(\frac{d}{\lambda}\theta_x\right). \tag{4.46}$$

As the small-scale fading follows the Nakagami-M distribution, it is easy to check that the signal power gain g_{x_0} is gamma distributed as $g_{x_0} \sim \mathrm{Gamma}(1, 1/M)$.

The interferer's channel gain g_x is the product of the gamma distributed small-scale fading gain $|\omega_x|^2$ and the directional antenna array gain $G_{\text{sinc}}\left(\frac{d}{\lambda}\theta_x\right)$. Due to the complicated form of the actual antenna pattern $G_{\text{act}}(\cdot)$, the exact distribution of the interferer's power gain g_x is difficult to derive. As mentioned in the last section, the sinc antenna pattern is an excellent approximation of the actual antenna pattern with better analytical tractability, so we propose to adopt it in the analysis of mm-wave ad hoc networks.

Note that, different from the first application in this chapter, where the interferer's power gain g_x is gamma distributed, the interferer's power gain is much more complicated distributed in mm-wave networks. Therefore, the results for gamma distributed g_x derived in Chap. 3 cannot be directly applied to this application. In the following, we follow the rule of thumb for applying the analytical framework presented in Chap. 3, i.e., Methodology 1. A detailed proof is provided for the first result, i.e., coverage analysis for mm-wave ad hoc networks, to illustrate the main steps in applying the framework. The proofs for the remaining results follow similar steps and are therefore diverted to the appendix.

As illustrated in Methodology 1 presented in Chap. 3, the main task to derive the coverage probability $p_c(\tau)$ is to determine the entries in the matrix $\mathbf{C}_M$. First, a unique property of the directional array gain with the sinc antenna pattern is presented to help derive the coverage probability.

Lemma 4.3 *For* $p \in \mathbb{Z}^+$,

$$\int_0^{\infty}\left(\frac{\sin x}{x}\right)^{2p}\mathrm{d}x = \frac{\pi}{2(2p-1)!}\left\langle{2p-1 \atop p-1}\right\rangle, \tag{4.47}$$

where $\left\langle{n \atop k}\right\rangle$ *are the Eulerian numbers, i.e.,* $\left\langle{n \atop k}\right\rangle = \sum_{j=0}^{k+1}(-1)^j\binom{n+1}{j}(k-j+1)^n$.

Proof The proof can be found in [28, Lemma 2].

Following Methodology 1, the first step is to calculate the Laplace transform $\mathscr{L}(s)$. With Lemma 4.3, we are able to derive the log-Laplace transform $\eta(s)$ in the following proposition.

Proposition 4.3 *The log-Laplace transform* $\eta(s)$ *of mm-wave ad hoc networks with the sinc antenna pattern is given by*

$$\begin{aligned}\eta(s) = &-\frac{\pi R^2\lambda_{\mathrm{b}}\bar{\lambda}}{\alpha d N_{\mathrm{t}}}\sum_{p=1}^{\infty}\frac{(-s)^p\left\langle{2p-1 \atop p-1}\right\rangle\Gamma(M+p)}{R^{\alpha p}(2p-1)!p!(p-\delta)\,\Gamma(M)M^p}\\ &+\frac{\delta s^{\delta}\lambda_{\mathrm{b}}\bar{\lambda}}{dN_{\mathrm{t}}}\Gamma(-\delta)\frac{\Gamma(M+\delta)}{\Gamma(M)M^{\delta}}\zeta - \frac{s\sigma^2}{\beta P_{\mathrm{t}}N_{\mathrm{t}}}.\end{aligned} \tag{4.48}$$

Proof According to (2.3), the log-Laplace transform is written by

$$
\begin{aligned}
\eta(s) &= -2\pi\lambda_{\mathrm{b}}\int_{0}^{R}\left(1-\mathbb{E}_g[\exp(-sgx^{-\alpha})]\right)x\mathrm{d}x \\
&= -2\pi\lambda_{\mathrm{b}}\mathbb{E}_g\left[\int_{0}^{R}\left[1-\exp(-sgx^{-\alpha})\right]x\mathrm{d}x\right] \\
&= -\pi\lambda_{\mathrm{b}}R^2-\pi\lambda_{\mathrm{b}}\mathbb{E}_g\left[(sg)^{\delta}\int_{sgR^{-\alpha}}^{\infty}e^{-t}\mathrm{d}t^{-\delta}\right] \\
&= -\pi\lambda_{\mathrm{b}}\left\{R^2-\delta R^2\mathbb{E}_g\left[\mathrm{E}_{1+\delta}(sR^{-\alpha}g)\right]\right\}.
\end{aligned} \tag{4.49}
$$

A part of the log-Laplace transform can be recast as

$$
\begin{aligned}
&R^2-\delta R^2\mathbb{E}_g\left[\mathrm{E}_{1+\delta}(sR^{-\alpha}g)\right] \\
&= R^2-\delta R^2\left\{\frac{s^{\delta}}{R^2}\Gamma(-\delta)\,\mathbb{E}_g\left[g^{\delta}\right]+\frac{\alpha}{2}-\sum_{p=1}^{\infty}\frac{(-s)^p}{R^{\alpha p}p!\,(p-\delta)}\mathbb{E}_g\left[g^p\right]\right\} \qquad (4.50) \\
&= \delta R^2\sum_{p=1}^{\infty}\frac{(-s)^p}{R^{\alpha p}p!\,(p-\delta)}\mathbb{E}_g\left[g^p\right]-\delta s^{\delta}\Gamma(-\delta)\,\mathbb{E}_g\left[g^{\delta}\right] \\
&= \delta R^2\sum_{p=1}^{\infty}\frac{(-s)^p\Gamma(M+p)}{R^{\alpha p}p!\,(p-\delta)\,\Gamma(M)M^p}\int_0^1\frac{\sin^{2k}\left(\frac{\pi d}{\lambda}N_{\mathrm{t}}\theta\right)}{\left(\frac{\pi d}{\lambda}N_{\mathrm{t}}\theta\right)^{2k}}\mathrm{d}\theta \\
&\quad -\delta s^{\delta}\Gamma(-\delta)\frac{\Gamma(M+\delta)}{\Gamma(M)M^{\delta}}\int_0^1\left|\frac{\sin\left(\frac{\pi d}{\lambda}N_{\mathrm{t}}\theta\right)}{\frac{\pi d}{\lambda}N_{\mathrm{t}}\theta}\right|^{2\delta}\mathrm{d}\theta \\
&\le \frac{R^2\bar{\lambda}}{\alpha dN_{\mathrm{t}}}\sum_{p=1}^{\infty}\frac{(-s)^p\binom{2p-1}{p-1}\Gamma(M+p)}{R^{\alpha p}(2p-1)!p!\,(p-\delta)\,\Gamma(M)M^p}-\frac{\delta s^{\delta}\bar{\lambda}}{\pi dN_{\mathrm{t}}}\Gamma(-\delta)\frac{\Gamma(M+\delta)}{\Gamma(M)M^{\delta}}\zeta,
\end{aligned} \tag{4.51}
$$

where (4.50) is from the series expansion of the generalized exponential integral, (4.51) follows Lemma 4.3, and the upper bound is derived by extending the integral upper limit to infinity given that, for the tiny ripple tails of the $2k$-th power of the sinc, the additional integration values are extremely small, and thus, the upper bound in (4.51) is tight. Therefore, the proof of Proposition 4.3 is complete.

The second step in Methodology 1 is to calculate the n-th ($1 \le n \le M-1$) derivatives of $\eta(s)$. Based on Proposition 4.3, it is easy to calculate $\{a_n\}_{n=0}^{M-1}$ and derive a lower bound of the coverage probability with the sinc antenna pattern, shown in the following proposition.

Proposition 4.4 *The coverage probability of mm-wave ad hoc networks with the sinc antenna pattern is tightly lower bounded by*

$$
p_{\mathrm{c}}^{\mathrm{sinc}}(\tau) \ge \left\|\exp\left(\frac{1}{N_{\mathrm{t}}}\mathbf{A}_M\right)\right\|_1. \tag{4.52}
$$

The nonzero coefficients in the lower triangular Toeplitz matrix $\mathbf{A}_M$ *are given by*

$$a_k = \frac{(-1)^{k+1}}{k!}\left[\frac{\pi R^2\lambda_\mathrm{b}\bar{\lambda}}{\alpha d}\sum_{p=\max\{1,k\}}^{\infty}\frac{(-\tau r_0^\alpha)^p\binom{2p-1}{p-1}\Gamma(M+p)}{R^{\alpha p}(2p-1)!(p-k)!\,(p-\delta)\,\Gamma(M)}\right.$$
$$\left.-\frac{\delta\lambda_\mathrm{b}\bar{\lambda}}{d}(\delta)_k\,\Gamma(-\delta)\frac{\Gamma(M+\delta)}{\Gamma(M)}\tau^\delta r_0^2\zeta+\mathbb{1}(k\le 1)\frac{\tau M r_0^\alpha\sigma^2}{\beta P_\mathrm{t}}\right], \tag{4.53}$$

where $(x)_n$ *represents the falling factorial, and*

$$\zeta = \int_0^\infty \left|\frac{\sin x}{x}\right|^{2\delta} \mathrm{d}x. \tag{4.54}$$

Remark 4.2 According to recent mm-wave channel measurements [20], the path loss exponent α is less than 3, which ensures the convergence of ζ.

Remark 4.3 Although the expressions in Proposition 4.4 involve a summation of infinitely many terms, it turns out that, in practical evaluation, the series converges quickly, and the high-order terms contribute little to the sum. Hence, using a finite number of terms is sufficient for numerical computation. In addition, ζ only depends on the path loss exponent α and can easily be evaluated numerically and offline. Overall, the expression in Proposition 4.4 is much easier to evaluate than existing results [27, 34] that contain multiple nested integrals.

Remark 4.4 For a given coverage probability, the maximum transmitter density can be numerically determined by Proposition 4.4.

Mm-Wave Cellular Networks

While the sinc antenna pattern can still be employed to get a highly accurate approximation of the actual antenna pattern, its numerical evaluation is more complicated and the expression reveals little insight. In particular, as (3.74) showed, an additional integral is needed over the distance between the serving BS and the typical user. Furthermore, since the minimum distance between the interfering BSs and the typical receiver is r_0 in cellular networks, i.e., $l(r_0) = r_0$, the summation of infinite terms in the integrand does not converge quickly. Instead, we analyze the cosine antenna pattern in this section, which provides a more tractable expression.

In contrast to existing works [24], we present an analytical result for the coverage probability that fully reflects the directionality in mm-wave cellular networks. Note that although the proposed approximate antenna pattern is more complicated than the flat-top pattern, the new expression based on the analytical framework in Chap. 3 is more compact and tractable. With the cosine antenna pattern, the coverage probability is derived in the following proposition.

Proposition 4.5 *The coverage probability of mm-wave cellular networks with the cosine antenna pattern is given by*

$$p_{\mathrm{c}}^{\cos}(\tau) = \pi\lambda_{\mathrm{b}} \int_0^{R^2} e^{-\pi\lambda_{\mathrm{b}} r} \left\| \exp\left\{ \frac{1}{N_{\mathrm{t}}} \mathbf{C}_M(r) \right\} \right\|_1 \mathrm{d}r. \tag{4.55}$$

The nonzero entries in $\mathbf{C}_M$ *are determined by*

$$\begin{aligned} c_k(r) = & \frac{2\sqrt{\pi}\lambda_{\mathrm{b}}\bar{\lambda}\Gamma\left(k+\frac{1}{2}\right)\Gamma(M+k)\tau^k}{d(k!)^2(\alpha k-2)\Gamma(M)} \left[J_k(-\tau) r - J_k\left(-\frac{\tau}{R^\alpha} r^{\frac{1}{\delta}}\right) R^{2-\alpha k} r^{\frac{k}{\delta}} \right] \\ & + \mathbb{1}(k \leq 1) \frac{(-1)^{k+1} M\tau\sigma^2}{\beta P_{\mathrm{t}}}, \end{aligned} \tag{4.56}$$

where

$$J_k(x) = {}_3F_2\left(k+\frac{1}{2}, k-\delta, k+M; k+1, k+1-\delta; x\right), \tag{4.57}$$

with ${}_3F_2(a_1, a_2, a_3; b_1, b_2; z)$ *denoting the generalized hypergeometric function.*

Proof See Appendix.

Note that the coefficients $c_k(r)$ in Proposition 4.5 can be expressed based on the well-known hypergeometric function rather than the infinite summations as in Proposition 4.4 for ad hoc networks, which are efficiently calculated in modern numerical software. This illustrates that the cosine antenna pattern enables a more tractable analysis for cellular networks. We use Proposition 4.4 as an approximation for the coverage probability with the actual antenna pattern, while the accuracy of the cosine antenna pattern will be verified later via simulation.

Remark 4.5 With Proposition 4.5, we numerically calculate the required BS density as well as the minimum number of antennas for a desirable coverage probability. Furthermore, the optimal BS density that achieves the maximum coverage probability can also be numerically determined by Proposition 4.5.

In Fig. 4.9a, the SINR coverage probabilities for mm-wave ad hoc networks are evaluated. It is observed that the analytical results match the simulations with negligible gaps, which implies the accuracy of the bound in Proposition 4.4. In Remark 4.3, we mentioned that using finite terms for the summations in $\{c_k\}_{k=0}^{M-1}$ is sufficient. In the numerical evaluation of Proposition 4.4 in Fig. 4.9a, we only use 5 terms in the summations, and it turns out that the higher order terms are negligible for practical evaluation.

In Fig. 4.9b, the coverage probability for a mm-wave cellular network is evaluated. We see that both the analytical results in Proposition 4.5 and Corollary 4.2 give an approximate coverage probability with minor gaps. The expression in Proposition 4.5 yields a very good approximation for smaller SINR thresholds and a tight bound for larger ones. This is because the major approximations made in the cosine antenna pattern (4.44) are on the side lobe gains that are approximated to be zeros, while the main lobe gains are approximated accurately with the cosine function. When the

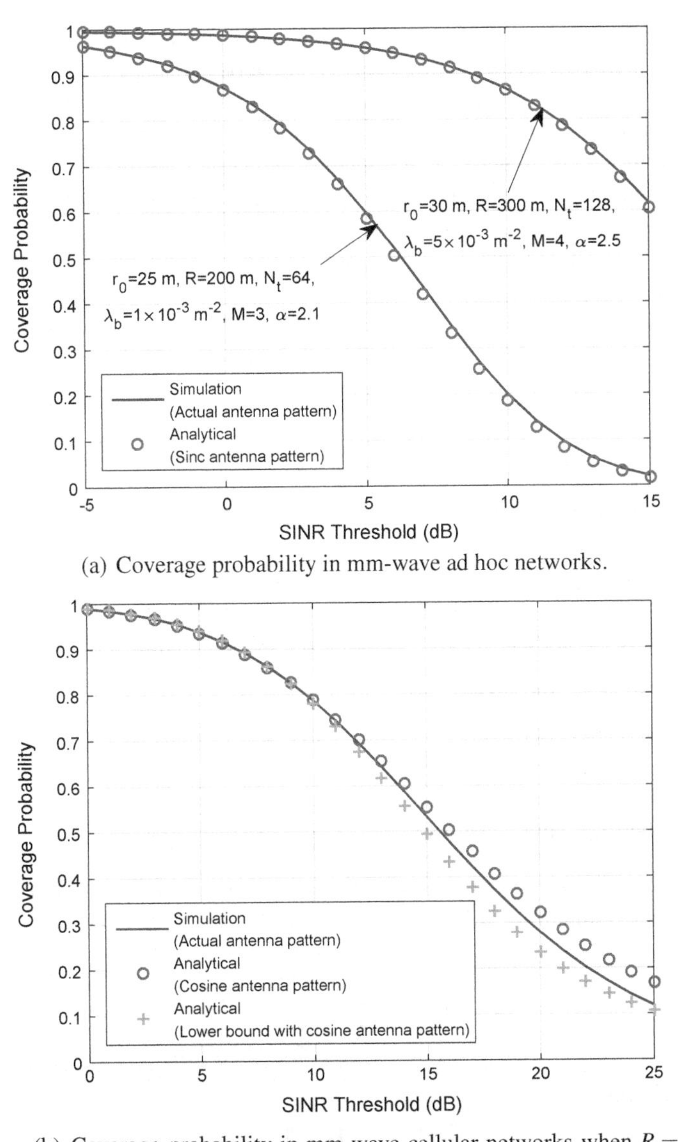

(a) Coverage probability in mm-wave ad hoc networks.

(b) Coverage probability in mm-wave cellular networks when $R = 200$ m, $N_t = 128$, $\lambda_b = 1 \times 10^{-3}$ m^{-2}, $M = 3$, and $\alpha = 2.1$.

Fig. 4.9 Coverage analysis using **a** Proposition 4.4 for mm-wave ad hoc networks and **b** Proposition 4.5 and Corollary 4.2 for mm-wave cellular networks. ©2017 IEEE. Reprinted, with permission, from [23]

SINR threshold gets large, the interference power is smaller, which also means the interference is more likely to be produced by side lobe gains. Therefore, the gap will gradually increase due to the relatively crude approximation of the side lobe gains.

The analytical result in Corollary 4.2 provides a lower bound of the expression in Proposition 4.5. Although it is not guaranteed to be a lower or upper bound of the exact SINR coverage probability, it gives a good approximation as shown in Fig. 4.9b, with more analytical tractability and potential for further analysis, which will be discussed in detail in the next subsection. The results presented in Fig. 4.9b show the effectiveness and rationale of the proposed cosine antenna pattern (4.44) in coverage analysis for mm-wave cellular networks, which is an ideal candidate for further performance analysis in mm-wave cellular networks.

4.4.3 Impact of Directional Antenna Arrays

How the directional antenna array size affects the performance of the network is a critical and unique problem in mm-wave systems. In this section, we investigate this effect with the analytical results of coverage probabilities derived in the last section that adopt more accurate approximations for the actual antenna pattern.

Mm-Wave Ad Hoc Networks

We first investigate how directional antenna arrays affect the coverage probability in mm-wave ad hoc networks. Generally speaking, increasing the array size enhances the signal quality, but may also increase interference power. The overall effect is revealed in the following corollary.

Corollary 4.1 *The tight lower bound of the coverage probability* (4.52) *is a nondecreasing concave function with the array size, and it can be rewritten as*

$$p_{\mathrm{c}}^{\mathrm{sinc}}(t) \geq e^{c_0 t}\left(1+\sum_{n=1}^{M-1}\beta_n t^n\right), \tag{4.58}$$

where $t = \frac{1}{N_{\mathrm{t}}}$, *and*

$$\beta_n = \frac{\left\|(\mathbf{C}_M - c_0\mathbf{I}_M)^n\right\|_1}{n!} \quad n \geq 1. \tag{4.59}$$

When $t \to 0$, *i.e.,* $N_{\mathrm{t}} \to \infty$, *the asymptotic outage probability is given by*

$$\tilde{p}_{\mathrm{o}}^{\mathrm{sinc}}(t) \sim \frac{\mu}{N_{\mathrm{t}}}, \tag{4.60}$$

where $\mu = -\sum_{n=0}^{M-1} c_n > 0$.

Proof See Appendix.

From Corollary 4.1, we see that $p_{\mathrm{c}}^{\mathrm{sinc}}(t) \to 1$ as $t \to 0$ for all network parameters. Hence, for any desired coverage requirement $1-\varepsilon$, there exists a minimum antenna array size N_{t} that can satisfy it regardless of the other network parameters, which can be numerically determined by Corollary 4.1. The lower bound in Corollary 4.1 indicates how antenna arrays affect the coverage probability. From Corollary 4.1, we discover that increasing the directional antenna array size will definitely benefit the coverage probability in ad hoc networks. Moreover, we see that the lower bound is a product of an exponential function and a polynomial function of order $M-1$ of the inverse of the array size t. For the special case that $M=1$, i.e., Rayleigh fading channel, the lower bound reduces to an exponential one. The asymptotic coverage probability (4.60) shows that the asymptotic outage probability is inversely proportional to the array size. To the best of the authors' knowledge, this is the first analytical result on the impact of antenna arrays in mm-wave network analysis.

Remark 4.6 Note that the manipulation in Corollary 4.1 is based on the analytical framework in Chap. 3. Especially, it benefits greatly from the delicate tackling of the gamma distributed signal power, via a lower triangular Toeplitz matrix representation. If the upper bound in [24, 27, 28, 34] was used instead, we would not be able explicitly to reveal the impact of antenna arrays, which, from another perspective, confirms the advantages of the analytical framework.

Mm-Wave Cellular Networks

In the last subsection, we have derived an analytical result for coverage probability of mm-wave cellular networks with the cosine antenna pattern. However, it is difficult to further analyze the impact of directional antenna arrays since there is an extra integral of the induced ℓ_1-norm of the matrix exponential, which contains the array size parameter N_{t}. As an alternative, a lower bound for the coverage probability in Proposition 4.5 is provided next, based on which we present the impact directional antenna arrays.

Corollary 4.2 *A lower bound of the coverage probability* (4.55) *is given by*

$$p_{\mathrm{c}}^{\cos}(\tau) \geq \left(1-e^{-\pi\lambda_{\mathrm{b}}R^2}\right)\left\|\exp\left\{\frac{1}{N_{\mathrm{t}}(1-e^{-\pi\lambda_{\mathrm{b}}R^2})}\mathbf{D}_M\right\}\right\|_1. \tag{4.61}$$

The nonzero entries in $\mathbf{D}_M$ *determined by*

$$\begin{aligned} d_k = & \frac{2\bar{\lambda}\Gamma\left(k+\frac{1}{2}\right)\Gamma(M+k)\tau^k}{\sqrt{\pi}d(k!)^2(\alpha k-2)\Gamma(M)}\Bigg[y_k(-\tau)-(\pi\lambda_{\mathrm{b}})^2R^{2-\alpha k} \\ & \times\int_0^{R^2} e^{-\pi\lambda_{\mathrm{b}}r}r^{\frac{\alpha k}{2}}J_k\left(-\frac{\tau}{R^\alpha}r^{\frac{1}{\delta}}\right)\mathrm{d}r\Bigg] \\ & +\mathbb{1}(k\leq 1)\frac{(-1)^{k+1}M\tau\sigma^2}{\beta P_{\mathrm{t}}(\pi\lambda_{\mathrm{b}})^{\frac{1}{\delta}}}\gamma\left(1+\frac{1}{\delta},\pi\lambda_{\mathrm{b}}R^2\right), \end{aligned} \tag{4.62}$$

where $\gamma(s, x)$ is the lower incomplete gamma function and

$$y_k(x) = J_k(x)\left[1 - e^{-\pi\lambda_b R^2}\left(1 + \pi\lambda_b R^2\right)\right] + \mathbb{1}(k=0)\left(\pi\lambda_b R^2 - 1 + e^{-\pi\lambda_b R^2}\right). \tag{4.63}$$

Proof See Appendix.

With this lower bound, the integrand no longer involves the induced ℓ_1-norm of a matrix exponential, which creates the possibility to disclose the impact of antenna arrays as stated in the following corollary.

Corollary 4.3 *The lower bound of the coverage probability (4.55) is a non-decreasing concave function of the array size, and it can be rewritten as*

$$p_c^{\cos}(t) \geq \left(1 - e^{-\pi\lambda_b R^2}\right) e^{\beta_0 t}\left(1 + \sum_{n=1}^{M-1} \beta_n t^n\right), \tag{4.64}$$

where $t = \frac{1}{N_t}$ is the inverse of the array size and

$$\beta_n = \begin{cases} \dfrac{d_0}{1 - e^{-\pi\lambda_b R^2}} & n = 0, \\ \dfrac{\left\|(\mathbf{D}_M - d_0\mathbf{I}_M)^n\right\|_1}{n!\left(1 - e^{-\pi\lambda_b R^2}\right)} & n \geq 1. \end{cases} \tag{4.65}$$

When $t \to 0$, i.e., $N_t \to \infty$, the asymptotic outage probability is given by

$$\tilde{p}_o^{\cos}(t) \sim \frac{\mu}{N_t} + e^{-\pi\lambda_b R^2}, \tag{4.66}$$

where $\mu = -\sum_{n=0}^{M-1} d_n > 0$.

Proof The proof is similar to that of Corollary 4.1.

It turns out that this lower bound of the coverage probability with the array size is quite similar to that in mm-wave ad hoc networks, yet with additional terms brought by the user association. This similarity shows that the impact of directional antenna arrays in mm-wave networks does not depend much on the user association strategy. Although this result is based on the cosine antenna pattern and a lower bound, later we will show its accuracy via simulations. Similar to Remark 4.6, the key tool here is the analytical framework presented in Chap. 3, which enables us to investigate the impact of antenna arrays in mm-wave cellular networks.

In the following, we discuss the impact of directional antenna arrays on coverage probability in mm-wave networks via numerical results.[7] Figure 4.10a demonstrates that the analytical result in Corollary 4.1 for mm-wave ad hoc networks well

[7]In Fig. 4.10a and b, the x-axes are reversed, and the y-axes are in the logarithm scale.

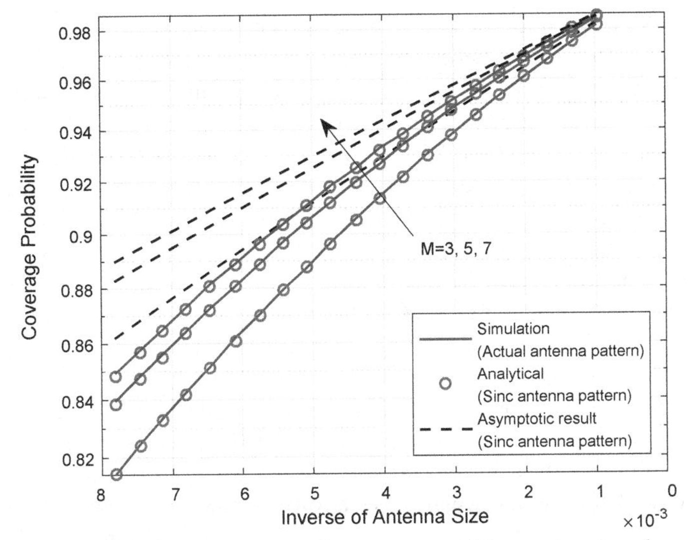

(a) Impact of antenna arrays in mm-wave ad hoc networks when $R = 200$ m, $\tau = 5$ dB, $\lambda_b = 1 \times 10^{-3}$ m^{-2}, $\alpha = 2.1$, and $r_0 = 25$ m.

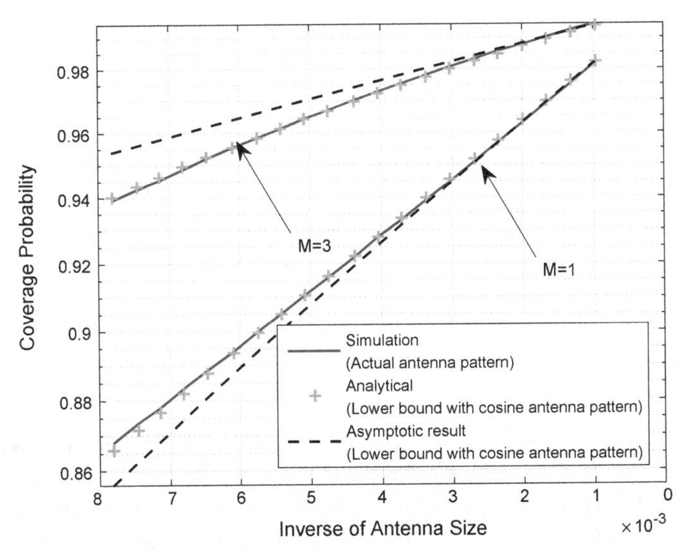

(b) Impact of antenna arrays in mm-wave cellular networks when $R = 200$ m, $\tau = 5$ dB, $\lambda_b = 1 \times 10^{-3}$ m^{-2}, and $\alpha = 2.1$.

Fig. 4.10 Investigation on the impact of antenna arrays using **a** Corollary 4.1 for mm-wave ad hoc networks and **b** Corollary 4.3 for mm-wave cellular networks. ©2017 IEEE. Reprinted, with permission, from [23]

matches the simulation result. We see that the increase of the array size leads to an improvement of the coverage probability, which confirms the monotonicity property in Corollary 4.1. In the following, we provide some intuitive explanations for this phenomenon. The increase of the array size increases the maximum array gain for both signal and interference at the same pace, in proportion to the array size, and therefore, there is almost no performance gain from increasing the maximum array gain via enlarging the array size. Nevertheless, another effect of the increasing array size for the interference is the narrowing of the beams, which reduces the probability that the interferers direct the main lobes toward the typical receiver. Moreover, note that the lower bound derived in Corollary 4.1 is non-decreasing concave, which means that the benefits on the coverage from leveraging more antennas gradually diminish with the increasing antenna size. In addition, we discover that the increase of the Nakagami parameter M results in an increase of the coverage probability.

In Fig. 4.10b, the impact of antenna arrays in mm-wave cellular networks is investigated. For the analytical result, we evaluate the coverage probability using the expression in Corollary 4.3, which gives a lower bound of the coverage probability adopting the cosine antenna pattern. Although Rayleigh fading, i.e., $M = 1$, is only a special case for the analysis in this paper and not suitable for LOS mm-wave channels, it is valuable to examine this special case for checking the lower bound in Corollary 4.3. When $M = 1$, the lower bound (4.64) reduces to an exponential one, which is linear in the logarithm scale shown in Fig. 4.10b. When the Nakagami parameter M increases, the polynomial term takes effect to make the lower bound to be a non-decreasing concave one. It turns out that the lower bound derived in Corollary 4.3 can serve as an effective expression for analyzing the impact of directional antenna arrays in mm-wave cellular networks, and that the cosine antenna pattern is a satisfactory surrogate of the actual antenna pattern for tractable analysis in mm-wave networks.

4.5 Summary

Two applications of the general analytical framework were presented in this chapter. In the first application, we investigated the effect of the BS density and antenna size on the network ASE and energy efficiency. In particular, we characterized the scaling of the network ASE with respect to the BS density and the number of BS antennas, respectively, and showed that there exists a phase transition phenomenon. Moreover, we derived key thresholds for identifying the effects of the BS density or the number of BS antennas on the energy efficiency.

In the second application, we considered mm-wave networks which bear complicated propagation characteristics. The coverage probabilities were derived for mm-wave ad hoc and cellular networks, where two approximate antenna patterns with good accuracy and analytical tractability were adopted. Based on the derived results, the impacts of directional antenna arrays in mm-wave networks were investigated. In particular, analytical results show that the coverage probabilities of both types of networks increase as a non-decreasing concave function with the antenna array size.

Bibliographical Notes

Compared with single-antenna cellular networks, there have been much fewer studies on the performance analysis of multi-antenna cellular networks. Chandrasekhar et al. (2009) analyzed the two-tier MIMO HetNets, with the first-order Taylor expansion applied to derive an approximate expression for the coverage probability. ASE optimization was investigated by Cheung et al. (2012) with approximations by ignoring the infinitesimals. Stochastic orders were introduced by Dhillon et al. (2013) to provide a qualitative comparison of different multi-antenna techniques. Soh et al. (2013) evaluated the energy efficiency of heterogeneous cellular networks based on the analytical expressions with complicated forms. ASE and energy efficiency were derived by Li et al. (2014) for MIMO cellular networks, with tractable expressions in the forms of Toeplitz matrices.

Millimeter-wave networks have drawn much attention in recent years. Analytical results for coverage and rate coverage probabilities in noise-limited mm-wave networks were presented by Singh et al. (2014). SINR and rate coverage based on a simplified directional antenna pattern were obtained for mm-wave cellular networks by Bai and Heath Jr. (2015). Lee et al. (2016) considered random beamforming in mm-wave networks and used the actual antenna pattern by adopting some asymptotic approximation in the analysis. Yu et al. (2017) proposed tractable approximated antenna patterns and investigated the impact of directional antenna arrays on the coverage probability, which provides the main results in Sect. 4.4.

Appendix

Proof of Lemma 4.1

Firstly, since $\mathbf{C}_M$ is a lower triangular Toeplitz matrix, $\bar{\mathbf{P}}_M$ is also a lower triangular Toeplitz matrix [13, 14], and the recurrence formula of $\bar{p}_n$ is given by [13, 14] as

$$\bar{p}_n = -c \sum_{i=0}^{n-1} \tilde{c}_{n-i} \bar{p}_i, \quad n \geq 1, \tag{4.67}$$

where $\bar{p}_0 = \frac{1}{1-p_\mathrm{a}+c_0}$ and $c = \frac{p_\mathrm{a}}{1-p_\mathrm{a}+c_0}$. Note that, according to (4.7), $k_0 > 0$ while $\tilde{c}_n < 0$ for $n \geq 1$. Therefore, it is easy to check the signs of $\{\bar{p}_n\}_{n=0}^{M-1}$.

Second, to prove the rest of the lemma, we define $\mathbf{B} \triangleq \left(\frac{1}{p_{\mathrm{a}}} - 1\right)\mathbf{I}_M + \tilde{\mathbf{C}}_M$. Then the derivative of $\left\|\bar{\mathbf{P}}_M\right\|_1$ w.r.t. p_{a} is given by

$$\frac{\partial \left\|\bar{\mathbf{P}}_M\right\|_1}{\partial p_{\mathrm{a}}} = \left\|\frac{\partial \bar{\mathbf{P}}_{M1}}{\partial p_{\mathrm{a}}}\right\|_1 = \left\|\frac{p_{\mathrm{a}}\frac{\partial \mathbf{B}^{-1}}{\partial p_{\mathrm{a}}} - \mathbf{B}^{-1}}{p_{\mathrm{a}}^2}\right\|_1, \tag{4.68}$$

Since $\frac{\partial \mathbf{B}^{-1}}{\partial p_{\mathrm{a}}} = -\mathbf{B}^{-1}\frac{\partial \mathbf{B}}{\partial p_{\mathrm{a}}}\mathbf{B}^{-1}$, we obtain $\frac{\partial \left\|\bar{\mathbf{P}}_M\right\|_1}{\partial p_{\mathrm{a}}} = \frac{1}{p_{\mathrm{a}}}\left(\left\|\bar{\mathbf{P}}_M^2\right\|_1 - \left\|\bar{\mathbf{P}}_M\right\|_1\right)$.

Third, to help derive the upper bound of $\left\|\bar{\mathbf{P}}_M\right\|_1$, we define

$$k_0 = \tilde{c}_0 - 1, \quad \mathbf{K}_M = \tilde{c}_0\mathbf{I}_M - \tilde{\mathbf{C}}_M. \tag{4.69}$$

Then, we rewrite $\left\|\bar{\mathbf{P}}_M\right\|_1$ as

$$\left\|\bar{\mathbf{P}}_M\right\|_1 = \frac{1}{p_{\mathrm{a}}}c\left\|(\mathbf{I} - c\mathbf{K}_M)^{-1}\right\|_1, \tag{4.70}$$

where $c = \frac{p_{\mathrm{a}}}{1+k_0 p_{\mathrm{a}}} = \frac{p_{\mathrm{a}}}{1-p_{\mathrm{a}}+c_0}$. Since $(\mathbf{I} - c\mathbf{K}_M)(\mathbf{I} - c\mathbf{K}_M)^{-1} = \mathbf{I}$, we have

$$(\mathbf{I} - c\mathbf{K}_M)^{-1} = \mathbf{I} + c\mathbf{K}_M(\mathbf{I} - c\mathbf{K}_M)^{-1}. \tag{4.71}$$

Using the triangle inequality, we obtain

$$\left\|[\mathbf{I} - c\mathbf{K}_M]^{-1}\right\|_1 \leq \|\mathbf{I}\|_1 + c\|\mathbf{K}_M\|_1\left\|[\mathbf{I} - c\mathbf{K}_M]^{-1}\right\|_1, \tag{4.72}$$

which can be written as $\left\|[\mathbf{I} - c\mathbf{K}_M]^{-1}\right\|_1 \leq \frac{\|\mathbf{I}\|_1}{1-c\|\mathbf{K}_M\|_1}$.

As $\|\mathbf{I}\|_1 = 1$, and $\|\mathbf{K}_M\|_1 = \sum_{i=1}^{M-1} k_i$, we get an upper bound of $\|\mathbf{T}_M\|_1$ as

$$\|\mathbf{T}_M\|_1 \leq \frac{1}{(1+k_0 p_{\mathrm{a}})\left(1 - c\sum_{i=1}^{M-1} k_i\right)} = \frac{1}{1 + p_{\mathrm{a}}\left(k_0 - \sum_{i=1}^{M-1} k_i\right)}. \tag{4.73}$$

For the lower bound, we define $\mathbf{x} \triangleq [1, 1, \ldots, 1]^T$ and $\mathbf{y} \triangleq \mathbf{B}\mathbf{x}$. As $\mathbf{B}$ is a non-singular matrix, then we have $\mathbf{x} = \mathbf{B}^{-1}\mathbf{y}$. Using the inequality $\|\mathbf{x}\|_1 \leq \left\|\mathbf{B}^{-1}\right\|_1\|\mathbf{y}\|_1$, we get $\left\|\mathbf{B}^{-1}\right\|_1 \geq \frac{\|\mathbf{x}\|_1}{\|\mathbf{y}\|_1}$. Since $\mathbf{y} = \mathbf{B}\mathbf{x}$, we get $\|\mathbf{y}\|_1 = M\left(k_0 + \frac{1}{p_{\mathrm{a}}}\right) - (M-1)k_1 - \cdots - k_{M-1}$. Therefore, we have the following lower bound of $\|\mathbf{T}_M\|_1$

$$\|\mathbf{T}_M\|_1 \geq \frac{1}{p_{\mathrm{a}}}\frac{\|\mathbf{x}\|_1}{\|\mathbf{y}\|_1} = \frac{1}{1 + p_{\mathrm{a}}\left(k_0 - \sum_{i=1}^{M-1} k_i + \sum_{i=1}^{M-1}\frac{i}{M}k_i\right)}. \tag{4.74}$$

Note that it can be shown that $k_i > k_{i+1}$ for $i \in \mathbb{N}$, and $\sum_{i=1}^{\infty} k_i = k_0$. Therefore, both $k_0 - \sum_{i=1}^{M-1}\left(1 - \frac{i}{M}\right)k_i$ and $k_0 - \sum_{i=1}^{M-1} k_i$ are positive, and the gap between them $\sum_{i=1}^{M-1} \frac{i}{M} k_i$ will be a decrease function with M when M is larger than a certain value.

Proof of Proposition 4.2

From (4.33), we find that it is not possible that the two inequalities $\xi_{\rm EE}(M) \leq \xi_{\rm EE}(M-1)$ and $\xi_{\rm EE}(M) \leq \xi_{\rm EE}(M+1)$ hold simultaneously, which implies that it would never happen that the energy efficiency first decreases and then increases as we keep increasing M. Moreover, we have $\lim_{M\to\infty} \xi_{\rm EE}(M) = 0$, and $\xi_{\rm EE}(1) > 0$. Considering all these facts, there can only be two different cases for the effect of M on the energy efficiency: (1) Energy efficiency decreases with M, so deploying a single antenna at each BS is more energy efficient than using multiple antennas; (2) Deploying multi-antenna BSs can achieve higher energy efficiency than single-antenna BSs and there is an optimal value of M.

To determine the optimal number of transmit antennas $M^\star$, we consider the inequalities

$$\begin{cases} \xi_{\rm EE}(M^\star) \geq \xi_{\rm EE}(M^\star - 1) \\ \xi_{\rm EE}(M^\star) \geq \xi_{\rm EE}(M^\star + 1). \end{cases}$$

Substituting (4.33) to the above inequalities, we find that the optimal $M^\star$ satisfies the condition

$$\frac{\sum_{n=0}^{M^\star-1} \bar{p}_n}{\bar{p}_{M^\star-1}} - M^\star \leq \frac{p_{\rm a}\left(\frac{1}{\xi}P_{\rm t}\right) + P_0}{p_{\rm a}P_{\rm c}} \leq \frac{\sum_{n=0}^{M^\star} \bar{p}_n}{\bar{p}_{M^\star}} - \left(M^\star + 1\right). \tag{4.75}$$

Define the function $F(M) \triangleq \frac{p_{\rm c}(M)}{\bar{p}_{M-1}} - M$, then the optimal $M^\star$ is the greatest integer that is smaller than the solution of (4.34).

Proof of Proposition 4.5

Following similar steps as in the proof of Proposition 4.3, part of the (4.49) can be derived as

$$\begin{aligned} &2\int_{r_0}^{R} (1 - \mathbb{E}_g[\exp(-sgx^{-\alpha})])x{\rm d}x \\ &= \delta R^2 \sum_{k=1}^{\infty} \frac{(-sR^{-\alpha})^k}{k!(k-\delta)} \mathbb{E}_g[g^k] - \delta r_0^2 \sum_{k=1}^{\infty} \frac{(-sr_0^{-\alpha})^k}{k!(k-\delta)} \mathbb{E}_g[g^k]. \end{aligned} \tag{4.76}$$

Based on the cosine antenna pattern (4.44), we have

$$\begin{aligned} &\sum_{k=1}^{\infty} \frac{(-z)^k}{k!(k-\delta)} \mathbb{E}_g[g^k] \\ &= \frac{\bar{\lambda}}{\pi d N_{\rm t}} \sum_{k=0}^{\infty} \frac{(-z)^k}{k!(k-\delta)} \int_0^{\pi} \cos^{2k}\frac{x}{2}{\rm d}x + \frac{\bar{\lambda}}{\delta d N_{\rm t}} \end{aligned}$$

$$= \frac{\bar{\lambda}}{\sqrt{\pi} d N_t} \sum_{k=0}^{\infty} \frac{(-z)^k \Gamma\left(\frac{1}{2}+k\right)}{(k!)^2(k-\delta)} + \frac{\bar{\lambda}}{\delta d N_t}$$

$$= \frac{\bar{\lambda}}{\delta d N_t}\left[1 - {}_3F_2\left(\frac{1}{2}, -\delta, M; 1, 1-\delta; -\frac{z}{M}\right)\right], \tag{4.77}$$

where (4.77) inversely applies the definition (series expansion) of the generalized hypergeometric function [38, pp. 1000].

Substituting (4.77) into (4.76), the exponent of the Laplace transform is given by

$$\eta(s) = -\frac{s\sigma^2}{\beta P_t N_t} - \frac{\pi \lambda_b \bar{\lambda}}{d N_t}\left\{\left[J_0\left(-\frac{s r_0^{-\alpha}}{M}\right) - 1\right] r_0^2 - \left[J_0\left(-\frac{s R^{-\alpha}}{M}\right) - 1\right] R^2\right\}. \tag{4.78}$$

Note that the derivative for the generalized hypergeometric function is

$$\begin{aligned} &\frac{\mathrm{d}}{\mathrm{d}z} {}_3F_2(a_1, a_2, a_3; b_1, b_2; z) \\ &= \frac{\prod_{i=1}^{3} a_i}{\prod_{j=1}^{2} b_j} {}_3F_2(a_1+1, a_2+1, a_3+1; b_1+1, b_2+1; z). \end{aligned} \tag{4.79}$$

Based on this expression and Theorem 3.2, the entries in $\mathbf{C}_M$ in Proposition 4.5 are obtained.

Proof of Corollary 4.1

According to (4.49), the Laplace transform of noise and interference is

$$\begin{aligned} \mathscr{L}(s) &= p_0 = \exp\{\eta(s)\} \\ &= \exp\left(-s\sigma_n^2 - \pi\lambda_b R^2\left\{1 - \delta\mathbb{E}_g\left[\mathrm{E}_{1+\delta}(sR^{-\alpha}g)\right]\right\}\right). \end{aligned} \tag{4.80}$$

Note that $1 - \delta\mathbb{E}_g\left[\mathrm{E}_{1+\delta}(sR^{-\alpha}g)\right]$ is a positive term due to the facts that $\mathrm{E}_{1+\delta}(z)$ is a monotone decreasing function of z and $\mathrm{E}_{1+\delta}(0) = \frac{1}{\delta}$. Hence, the Laplace transform p_0 is non-decreasing with the antenna array size N_t, where $\eta(s)$ is given in (4.48). According to the recursive relationship (3.58) between p_n, it turns out that every p_n is a non-decreasing function of N_t. Recalling that $p_c(\tau) = \mathbb{E}_{r_0}\left[\sum_{n=0}^{M-1} p_n\right]$, the monotonicity in Corollary 4.1 has been proved, and the concavity of the lower bound can be proved via similar steps.

We first write $\mathbf{A}_M$ in the form

$$\mathbf{A}_M = a_0\mathbf{I}_M + (\mathbf{A}_M - a_0\mathbf{I}_M), \tag{4.81}$$

where the first term is a scalar matrix. Since $\mathbf{A}_M$ is a lower triangular Toeplitz matrix, the second part is a nilpotent matrix, i.e., $(\mathbf{A}_M - a_0\mathbf{I}_M)^n = \mathbf{0}$ for $n \geq M$. Hence, according to the properties of matrix exponential, we have

$$\exp\left\{\frac{1}{N_{\mathrm{t}}}\mathbf{A}_M\right\} = e^{a_0\frac{1}{N_{\mathrm{t}}}}\sum_{n=0}^{M-1}\frac{1}{n!}\left[\frac{1}{N_{\mathrm{t}}}\left(\mathbf{A}_M - a_0\mathbf{I}_M\right)\right]^n. \tag{4.82}$$

Since Theorem 3.2 has shown that $a_k > 0$ for $k \geq 1$, $\mathbf{A}_M - a_0\mathbf{I}_M$ is a strictly lower triangular Toeplitz matrix with all positive entries, and so are the matrices $(\mathbf{A}_M - a_0\mathbf{I}_M)^n$. Therefore,

$$\left\|\exp\left\{\frac{1}{N_{\mathrm{t}}}\mathbf{A}_M\right\}\right\|_1 = e^{a_0\frac{1}{N_{\mathrm{t}}}}\sum_{n=0}^{M-1}\frac{1}{n!}\left[\frac{1}{N_{\mathrm{t}}^n}\left\|(\mathbf{A}_M - a_0\mathbf{I}_M)^n\right\|_1\right], \tag{4.83}$$

which completes the proof of Corollary (4.58). When $t \to 0$, by omitting the higher order terms, the linear Taylor expansion of the coverage is

$$\left\|\exp\left\{\frac{1}{N_{\mathrm{t}}}\mathbf{A}_M\right\}\right\|_1 \sim 1 + \frac{a_0 + \|\mathbf{A}_M - a_0\mathbf{I}_M\|_1}{N_{\mathrm{t}}} = 1 + \frac{\sum_{n=0}^{M-1} a_n}{N_{\mathrm{t}}}, \tag{4.84}$$

where the slope satisfies

$$\sum_{n=0}^{M-1} a_n \overset{(B.\mathrm{I})}{<} \sum_{n=0}^{\infty} a_n = \sum_{n=0}^{\infty}\frac{(-s)^n}{n!}\eta^{(n)}(s) \overset{(B.\mathrm{II})}{=} \eta(0) = 0. \tag{4.85}$$

Step (B.I) follows the fact that $a_k > 0$ for $k \geq 1$, and (B.II) follows from the Taylor expansion of $\eta(0)$ at point s.

Proof of Corollary 4.2

We adopt the power series in (3.95) to help prove the corollary.

$$\bar{P}(z) \triangleq \mathbb{E}_{r_0}\left[P(z)\right] = \sum_{n=0}^{\infty}\bar{p}_n z^n. \tag{4.86}$$

Recall that $X(z) = \exp\{C(z)\}$, and we obtain the following lower bound with a slight abuse of notation due to the fact that $C(z)$ is a function of r_0 in cellular networks:

$$\begin{aligned}
\bar{P}(z) &= \pi\lambda_{\mathrm{b}}\int_0^{R^2}\exp\left\{-\pi\lambda_{\mathrm{b}}r + \frac{1}{N_{\mathrm{t}}}C(z;r)\right\}\mathrm{d}r \\
&= \pi\lambda_{\mathrm{b}}\int_0^{R^2} e^{-\pi\lambda_{\mathrm{b}}r}\exp\left\{\frac{1}{N_{\mathrm{t}}}\sum_{k=0}^{\infty}c_k(r)z^k\right\}\mathrm{d}r \\
&\geq \left(1 - e^{-\pi\lambda_{\mathrm{b}}R^2}\right)\exp\left\{\frac{\pi\lambda_{\mathrm{b}}}{N_{\mathrm{t}}(1 - e^{-\pi\lambda_{\mathrm{b}}R^2})}\sum_{k=0}^{\infty}\left(\int_0^{R^2} e^{-\pi\lambda_{\mathrm{b}}r}c_k(r)\mathrm{d}r\right)z^k\right\} \\
&\triangleq \left(1 - e^{-\pi\lambda_{\mathrm{b}}R^2}\right)\exp\left\{\frac{1}{N_{\mathrm{t}}(1 - e^{-\pi\lambda_{\mathrm{b}}R^2})}D(z;r)\right\}.
\end{aligned} \tag{4.87}$$

In fact, $\bar{P}(z)$ can be viewed as $\left(1-e^{-\pi\lambda_{\rm b}R^2}\right)\mathbb{E}_{r_0'}\left[\exp\left\{\frac{1}{N_{\rm t}}C(z;r_0')\right\}\right]$ for the random variable with pdf $f_{r_0'}(r)=\frac{\pi\lambda_{\rm b}}{1-e^{-\pi\lambda_{\rm b}R^2}}e^{-\pi\lambda_{\rm b}r}$. Due to the convexity of the exponential function, we apply Jensen's inequality in (4.87) and obtain the lower bound. Therefore, the coverage probability is given by

$$p_{\rm c}^{\cos}(\tau)=\sum_{n=0}^{M-1}\frac{1}{n!}\frac{{\rm d}^n}{{\rm d}z^n}\bar{P}(z)\Big|_{z=0}, \tag{4.88}$$

which can be further expressed as in Corollary 4.3.

References

1. H.S. Dhillon, R.K. Ganti, F. Baccelli, J.G. Andrews, Modeling and analysis of K-tier downlink heterogeneous cellular networks. IEEE J. Sel. Areas Commun. **30**, 550–560 (2012)
2. J. Zhang, J. Andrews, Distributed antenna systems with randomness. IEEE Trans. Wirel. Commun. **7**, 3636–3646 (2008)
3. C. Lee, M. Haenggi, Interference and outage in poisson cognitive networks. IEEE Trans. Wirel. Commun. **11**, 1392–1401 (2012)
4. J. Ferenc, Z. Néda, On the size distribution of Poisson Voronoi cells. Phys. A Stat. Mech. Appl. **385**(2), 518–526 (2007)
5. T. Kiang, Random fragmentation in two and three dimensions. Zeitschrift fur Astrophysik **64**, 433 (1966)
6. D. Cao, S. Zhou, Z. Niu, Optimal base station density for energy-efficient heterogeneous cellular networks, in *Proceedings of IEEE International Conference on Communications (ICC)*, Ottawa, Canada (2012)
7. M. Kountouris, J.G. Andrews, Downlink SDMA with limited feedback in interference-limited wireless networks. IEEE Trans. Wirel. Commun. **11**, 2730–2741 (2012)
8. J.G. Andrews, F. Baccelli, R.K. Ganti, A tractable approach to coverage and rate in cellular networks. IEEE Trans. Commun. **59**, 3122–3134 (2011)
9. A.M. Hunter, J.G. Andrews, S. Weber, Transmission capacity of ad hoc networks with spatial diversity. IEEE Trans. Wirel. Commun. **7**, 5058–5071 (2008)
10. R.H.Y. Louie, M.R. McKay, I.B. Collings, Open-loop spatial multiplexing and diversity communications in ad hoc networks. IEEE Trans. Inf. Theory **57**, 317–344 (2011)
11. Y. Wu, R.H.Y. Louie, M.R. McKay, I.B. Collings, Generalized framework for the analysis of linear MIMO transmission schemes in decentralized wireless ad hoc networks. IEEE Trans. Wirel. Commun. **11**, 2815–2827 (2012)
12. M. Kountouris, J.G. Andrews, Transmission capacity scaling of SDMA in wireless ad hoc networks, in *Proceedings of 2009 IEEE Information Theory Workshop*, Volos, Greece (2009), pp. 534–538
13. A. Vecchio, A bound for the inverse of a lower triangular Toeplitz matrix. SIAM J. Matrix Anal. Appl. **24**(4), 1167–1174 (2003)
14. D. Commenges, M. Monsion, Fast inversion of triangular Toeplitz matrices. IEEE Trans. Autom. Control **29**(3), 250–251 (1984)
15. J. Andrews, H. Claussen, M. Dohler, S. Rangan, M. Reed, Femtocells: past, present, and future. IEEE J. Sel. Areas Commun. **30**, 497–508 (2012)
16. C. Li, J. Zhang, K.B. Letaief, Throughput and energy efficiency analysis of small cell networks with multi-antenna base stations. IEEE Trans. Wirel. Commun. **13**, 2505–2517 (2014)

17. Z. Hasan, H. Boostanimehr, V.K. Bhargava, Green cellular networks: a survey, some research issues and challenges. IEEE Commun. Surv. Tutor. **13**, 524–540 (2011)
18. G. Auer, V. Giannini, C. Desset, I. Godor, P. Skillermark, M. Olsson, M.A. Imran, D. Sabella, M.J. Gonzalez, O. Blume, A. Fehske, How much energy is needed to run a wireless network? IEEE Wirel. Commun. **18**, 40–49 (2011)
19. G. Auer et al., D2.3: energy efficiency analysis of the reference systems, areas of improvements and target breakdown, *INFSO-ICT-247733 EARTH (Energy Aware Radio and NeTwork TecHnologies)* (2010)
20. M.R. Akdeniz, Y. Liu, M.K. Samimi, S. Sun, S. Rangan, T.S. Rappaport, E. Erkip, Millimeter wave channel modeling and cellular capacity evaluation. IEEE J. Sel. Areas Commun. **32**, 1164–1179 (2014)
21. T.S. Rappaport, R.W. Heath Jr., R.C. Daniels, J.N. Murdock, *Millimeter Wave Wireless Communications* (Pearson Education, 2014)
22. M. Haenggi, *Stochastic Geometry for Wireless Networks* (Cambridge University Press, Cambridge, U.K., 2012)
23. X. Yu, J. Zhang, M. Haenggi, K.B. Letaief, Coverage analysis for millimeter wave networks: the impact of directional antenna arrays. IEEE J. Sel. Areas Commun. **35**, 1498–1512 (2017)
24. T. Bai, R.W. Heath Jr., Coverage and rate analysis for millimeter-wave cellular networks. IEEE Trans. Wirel. Commun. **14**, 1100–1114 (2015)
25. X. Zhang, J.G. Andrews, Downlink cellular network analysis with multi-slope path loss models. IEEE Trans. Commun. **63**, 1881–1894 (2015)
26. J.G. Andrews, T. Bai, M.N. Kulkarni, A. Alkhateeb, A.K. Gupta, R.W. Heath Jr., Modeling and analyzing millimeter wave cellular systems. IEEE Trans. Commun. **65**, 403–430 (2017)
27. A. Thornburg, T. Bai, R.W. Heath Jr., Performance analysis of outdoor mmWave ad hoc networks. IEEE Trans. Signal Process. **64**, 4065–4079 (2016)
28. X. Yu, J. Zhang, K.B. Letaief, Coverage analysis for dense millimeter wave cellular networks: the impact of array size, in *IEEE Wireless Communications and Networking Conference* (2016), pp. 1–6
29. G. Lee, Y. Sung, J. Seo, Randomly-directional beamforming in millimeter-wave multiuser MISO downlink. IEEE Trans. Wirel. Commun. **15**, 1086–1100 (2016)
30. G. Lee, Y. Sung, M. Kountouris, On the performance of random beamforming in sparse millimeter wave channels. IEEE J. Sel. Top. Signal Process. **10**, 560–575 (2016)
31. A. Alkhateeb, G. Leus, R.W. Heath Jr., Limited feedback hybrid precoding for multi-user millimeter wave systems. IEEE Trans. Wirel. Commun. **14**, 6481–6494 (2015)
32. J. Brady, N. Behdad, A.M. Sayeed, Beamspace MIMO for millimeter-wave communications: system architecture, modeling, analysis, and measurements. IEEE Trans. Antennas Propag. **61**, 3814–3827 (2013)
33. C.A. Balanis, *Antenna Theory: Analysis and Design* (Wiley, Hoboken, NJ, USA, 2005)
34. K. Venugopal, M.C. Valenti, R.W. Heath, Jr., Interference in finite-sized highly dense millimeter wave networks, in *Proceedings of Information Theory and Application (ITA)*, San Diego, CA, USA (2015), pp. 175–180
35. M.D. Renzo, Stochastic geometry modeling and analysis of multi-tier millimeter wave cellular networks. IEEE Trans. Wirel. Commun. **14**, 5038–5057 (2015)
36. S.L. Cotton, W.G. Scanlon, B.K. Madahar, Millimeter-wave soldier-to-soldier communications for covert battlefield operations. IEEE Commun. Mag. **47**, 72–81 (2009)
37. J. Qiao, X.S. Shen, J.W. Mark, Q. Shen, Y. He, L. Lei, Enabling device-to-device communications in millimeter-wave 5G cellular networks. IEEE Commun. Mag. **53**, 209–215 (2015)
38. D. Zwillinger, *Table of Integrals, Series, and Products* (Elsevier, Amsterdam, Netherlands, 2014)

Chapter 5
Optimization of Multi-Antenna Wireless Networks

Abstract This chapter presents two applications of the general analytical framework on network design and optimization. The first application considers interference coordination. A tractable expression is first derived for the coverage probability of a user-centric interference nulling strategy, which then helps to effectively optimize the interference coordination range. The presented interference nulling strategy can achieve performance gains about 35–40% compared with the non-coordination case. The second application studies general multiuser MIMO heterogeneous networks (HetNets). Exact and asymptotic expressions of the coverage probabilities are presented, which reveal that the SIR invariance property of SISO HetNets does not hold for MIMO HetNets. Instead, the coverage probability may decrease as the network density increases. It is proved that the maximum coverage probability is achieved by activating only one tier of BSs, while the maximum ASE is achieved by activating all the BSs. This reveals a unique trade-off between the ASE and link reliability in multiuser MIMO HetNets. To achieve the maximum ASE while guaranteeing a certain link reliability, efficient algorithms are provided to find the optimal BS densities.

5.1 Interference Management

Due to the broadcast nature of wireless communications and the reuse of the same frequency channel by multiple communication pairs, intercell interference is a bottleneck for the performance of wireless networks. Without effective interference management, the performance of mobile users will be severely degraded by the interfering co-channel transmissions from nearby BSs. The situation would be even worse for the upcoming 5G networks, since it has been proposed that 5G networks would have a dense deployment of BSs. It has been pointed out that without interference management, the outage probability of the typical transmission link will be higher than 40% in the cellular network with single-antenna BSs [1]. One common way to decrease the outage is to deploy multiple antennas. Nevertheless, without an effective interference management scheme, the outage probability could still be very high for high-speed transmissions.

X. Yu et al., *Stochastic Geometry Analysis of Multi-Antenna Wireless Networks*, https://doi.org/10.1007/978-981-13-5880-7_5

In recent years, multi-cell cooperation has been proposed as an effective way to mitigate interference. There are different types of cooperation strategies by assigning different temporal/spectral/spatial dimensions to users among different cells [2–7]. For example, nearby BSs can use different frequency bands to serve their users, so that they do not interfere with each other. Specifically, if three nearby BSs coordinate and use three different frequency bands, then the user served by one BS will not receive interference from the other two nearby BSs. However, the trade-off of this method is that each BS only uses 1/3 bandwidth to transmit signals, which decreases the throughput.

By utilizing multi-antenna techniques, more effective interference management methods have been developed. There are two main kinds of approaches. One is called *BS cooperation* or joint transmission. It requires the cooperated BSs to share user data and channel state information (CSI). The cooperated BSs act together to serve their users. This approach can achieve the best performance with, however, extremely high signaling overhead. The other approach is called *BS coordination* or interference nulling, which requires the coordinated BSs to exchange the control information only. In this approach, one BS transmits data to its user, while all the other coordinated BSs suppress interference for this user. Both approaches have been well studied in small-scale networks [3, 8, 9], but their effectiveness in densely deployed networks requires further investigation, since there are new features to consider at the network level. First, studies in small-scale networks consider only a few BSs and assume they all cooperate with each other. But in large-scale networks, it is impossible to cooperate all the BSs. Thus, it is unavoidable for users to receive interference from uncooperative BSs. Therefore, how to select BSs to cooperate for a particular user is a critical issue which requires delicate consideration. Second, in reality, BSs are deployed irregularly depending on the environment. How to effectively model the irregularity of the BS positions and how to design interference management due to such features are still unclear.

There have been a few attempts applying stochastic geometry to analyze wireless networks with interference management [10–15]. Many of them relied on coarse approximations, producing results either in complicated forms or not accurate for practical guidance. In this section, we present a *user-centric intercell interference nulling* strategy, which applies the analytical framework presented in Chap. 3. A tractable expression will be provided for the coverage probability, which will then be applied for performance optimization.

5.1.1 Interference Nulling Techniques

Interference nulling (IN) techniques are a set of interference management methods that rely on zero-forcing (ZF) beamforming, which was introduced in Sect. 3.2. Consider a small-scale network in a region that contains N BSs, and each BS has N_{t} antennas. Assume that each BS serves one user and cancels the interference to the other users in the region with ZF beamforming. As discussed in Sect. 3.2.2, to enable the ZF beamforming at the transmitter side for MU-MIMO systems, it requires $N_t \geq N$. It means that if an N_t-antenna BS serves one user, then it can null interference for at most $N_t - 1$ users.

In large-scale networks, it is impossible to coordinate all the BSs. A straightforward idea is to divide BSs into different *clusters*. In each cluster, BSs are coordinated, and users do not receive interference from the BSs in the same cluster. However, each user still receives interference from the BSs outside the cluster. For the users located at the edge of the cluster, such cluster-based interference nulling is ineffective. To overcome this issue, *user-centric* approaches have been proposed. One idea is that each user requests N nearest interfering BSs to null interference. In this monograph, we call this method as *fixed-number-based IN*. The advantage of the user-centric approach is that there is no edge-user, and all the users enjoy a similar performance. However, the fixed-number-based IN has a major disadvantage. Due to the irregular locations of the BSs, IN with a fixed-number N cannot guarantee to effectively suppress interference, and sometimes it is also unrealistic. On one side, if the user is surrounded by many BSs, then the $(N+1)$-th nearest BS may still generate significant interference to this user. In this case, this approach is not effective. On the other side, if the user is surrounded by a few BSs, then the N-th BS may be far away, and cannot and does not need to coordinate. To ask the very far away BSs to coordinate leads to a waste of resources. Overall, fixed-number-based IN is not an effective approach for interference management.

To seek an effective user-centric approach for interference management, it is essential to identify and suppress the *dominant interference* in the network for each user. This is the key element of the method to be presented next, where each served user will selectively request a subset of interfering BSs for interference nulling. An interfering BS will be in this subset if the ratio of the average power received from this interfering BS to the average power received from the home BS is larger than a certain threshold, i.e., if its interference is strong relative to the user's information signal power. Since each BS uses the same transmit power, the coordinating BSs can be determined by the relative distances to the interfering BSs and the home BS. Specifically, considering the typical user with distance r_0 to its home BS, i.e., $r_0 = \|x_0\|$, it will request all the interfering BSs within distance μr_0 (where $\mu \geq 1$) for interference nulling. In the following, we will call μr_0 as the *IN range* and the parameter μ as the *IN range coefficient*. Note that as the distance information is relatively easy to obtain, this approach to determine the IN range for each user incurs much less overhead than the ones based on instantaneous channel information. Moreover, as only the dominant interfering sources will be suppressed, it will lead to a more efficient utilization of the available radio resources, and better performance will thus be achieved.

Note that once determined, the value of μ is the same for all the users, i.e., this strategy has a single design parameter for the network. However, due to the random locations of BSs and users, the signal transmission distance r_0 is different in different cells, which means that the area of interference coordination regions will be different for different users. Figure 5.1 illustrates the BSs who will receive requests from the typical user, and all of them are within the annulus (the gray area) from radius r_0 to

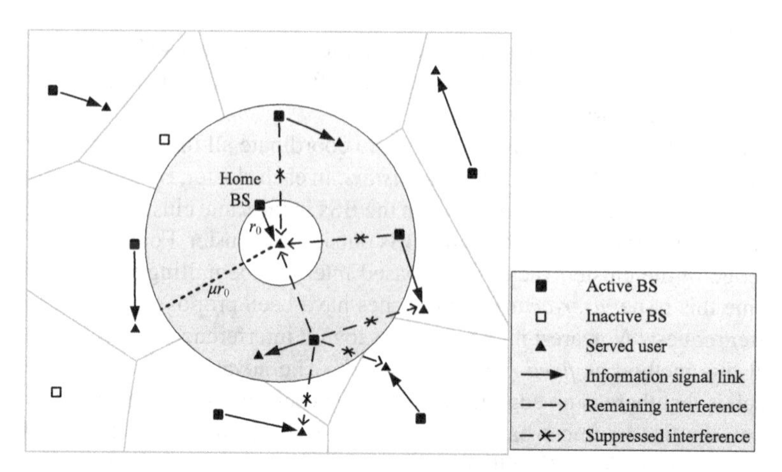

Fig. 5.1 A sample network where BSs and users are distributed as two independent PPPs. The typical user is located at the origin, and the interfering BSs in the gray region will receive coordination requests from the typical user, but some of them may not be able to suppress interference due to the degrees of freedom constraint. ©2015 IEEE. Reprinted, with permission, from [16]

μr_0. Thus, the number of coordination requests received by a BS is a random variable, i.e., a BS may belong to multiple annuluses centered around different users. Denote the number of requests received by the BS located at coordinate x as K_x. As K_x is random and unbounded, it is possible that $K_x \geq N_t$. Due to the limited spatial degrees of freedom, each BS can handle at most $N_t - 1$ requests. If a BS receives $K_x \geq N_t$ requests, we assume it will randomly choose $N_t - 1$ users to suppress interference. This implies that it is possible for the requesting user to receive interference from the BSs within the annulus (as shown in Fig. 5.1). Note that such phenomenon exists in many user-centric approaches. For example, for the fixed-number-based IN, the BS may also receive requests more than it can handle.

Remark 5.1 (*The effect of the IN range coefficient*) Tuning the parameter μ has conflicting effects: Increasing μ can suppress more nearby intercell interference. But the BSs will have less degrees of freedom for their own signal links, which will reduce the received information signal power. As a special case, $\mu = 1$ implies a *non-coordination scenario*, i.e., no interference nulling is employed in the network, and each active BS will serve its own user by single-user beamforming. The objective is to analytically evaluate the performance of this coordination strategy and find the optimal μ to achieve the best performance.

5.1.2 Network Model and Notations

We adopt the same network model as Sect. 4.1. The sets of BSs and users are denoted as Φ_b and Φ_u with densities λ_b and λ_u, respectively. Each BS serves one user at each time slot, and if the cell does not have any active user, this BS will not transmit and is called an inactive BS. Denote the set of active BSs as Φ'_b and the set of served users as Φ'_u. For the typical user located at the origin o, served by its home BS at location x_0, it will receive interference from the BSs outside the annulus, and probably also from the BSs within the annulus. In this way, the equivalent point process of the interfering BSs Φ' is composed of two parts. Let $\Phi_\mathrm{b}^{(1)}$ denote the set of interfering BSs farther than μr_0. Let $\Phi_\mathrm{b}^{(2)}$ denote the set of BSs who receive the request from the typical user but are unable to mitigate interference for this user. In addition, we assume Rayleigh fading channels. Then, according to (3.1), the received signal of this user is given by

$$\hat{s}_{x_0} = \sqrt{P_\mathrm{t} r_0^{-\frac{\alpha}{2}}} \mathbf{h}_{x_0 x_0} \mathbf{f}_{x_0} s_{x_0} + \sum_{x \in \Phi' = \Phi_\mathrm{b}^{(1)} \cup \Phi_\mathrm{b}^{(2)}} \sqrt{P_\mathrm{t}} \, \|x\|^{-\frac{\alpha}{2}} \mathbf{h}_{x_0 x} \mathbf{f}_x s_x + n_{x_0}, \tag{5.1}$$

where $\mathbf{f}_x$ is the $N_\mathrm{t} \times 1$ beamforming vector for the BS located at x.

Note that the typical user's home BS receives K_{x_0} requests, and thus, this BS will help $\min\left(K_{x_0}, N_\mathrm{t} - 1\right) \triangleq N_\mathrm{t} - M$[1] users to suppress interference. Denoting the channels of those requested users as $\mathbf{h}_{1x_0}, \ldots, \mathbf{h}_{(N_\mathrm{t}-M)x_0}$. According to (3.31) and (3.32), the ZF beamforming matrix is given by

$$\mathbf{F}_{x_0} = \mathbf{H}_{x_0}^{\dagger} = \mathbf{H}_{x_0}^H \left(\mathbf{H}_{x_0} \mathbf{H}_{x_0}^H\right)^{-1}, \tag{5.2}$$

where

$$\begin{aligned} \mathbf{F}_{x_0} &= \left[\mathbf{f}_{x_0}, \mathbf{f}_1, \ldots, \mathbf{f}_{N_\mathrm{t}-M}\right], \\ \mathbf{H}_{xx} &= \left[\mathbf{h}_{x_0 x_0}^T, \mathbf{h}_{1x_0}^T, \ldots, \mathbf{h}_{(N_\mathrm{t}-M)x_0}^T\right]^T. \end{aligned} \tag{5.3}$$

Therefore, by rewriting (5.2), the ZF beamforming vector for the typical user is given by

$$\mathbf{f}_{x_0} = \frac{\left(\mathbf{I}_{N_\mathrm{t}} - \mathbf{H}_{xx\backslash x_0}^H \left(\mathbf{H}_{xx\backslash x_0} \mathbf{H}_{xx\backslash x_0}^H\right)^{-1} \mathbf{H}_{xx\backslash x_0}\right) \mathbf{h}_{x_0 x_0}^H}{\left\| \left(\mathbf{I}_{N_\mathrm{t}} - \mathbf{H}_{xx\backslash x_0}^H \left(\mathbf{H}_{xx\backslash x_0} \mathbf{H}_{xx\backslash x_0}^H\right)^{-1} \mathbf{H}_{xx\backslash x_0}\right) \mathbf{h}_{x_0 x_0}^H \right\|}, \tag{5.4}$$

where $\mathbf{H}_{xx\backslash x_0} = \left[\mathbf{h}_{1x_0}^T, \ldots, \mathbf{h}_{(N_\mathrm{t}-M)x_0}^T\right]^T$.

As illustrated in Methodology 1 in Sect. 3.2.2, the first step to apply the general analytical framework is to calculate the conditional Laplace transform $\mathscr{L}(s)$. In order to derive it, we need to know the distributions of the signal power gain g_{x_0} and the interferer's power gain g_x, given in (3.16) and (3.17), which are correspondingly determined by [17]

---

[1] In Chap. 3, the shape parameter of the gamma distribution was denoted by M. Nevertheless, with a slight abuse of notation, we use $N_\mathrm{t} - M$ to denote the number of users that are helped by the home BS to null interference. Later we shall see that these parameters are closely related to each other.

$$g_{x_0} = \left|\mathbf{h}_{x_0}\mathbf{f}_{x_0}\right|^2 \sim \text{Gamma}\left(M, 1\right), \tag{5.5}$$

where $M = \max\left(N_\text{t} - K_{x_0}, 1\right)$, and

$$g_x = \left|\mathbf{h}_x^H \mathbf{f}_x\right|^2 \sim \text{Exp}\left(1\right). \tag{5.6}$$

Note that although the distributions of g_{x_0} and g_x are the same as those in (4.4) and (4.5), the spatial distribution of the interfering BSs, i.e., the point process Φ' is much more complicated in this part, and therefore forms one main challenge in performance analysis.

For the typical user, the distribution of receive SINR depends on K_{x_0}, i.e., the number of coordination requests received by its home BS. Thus, we denote the coverage probability of the user whose home BS receives k requests as

$$p_\text{c}\left(k\right) = \mathbb{P}\left(\text{SINR} \geq \tau \mid K_{x_0} = k\right). \tag{5.7}$$

Therefore, the average performance of the typical user is given by

$$p_\text{c} = \mathbb{E}_{K_{x_0}}\left[p_\text{c}\left(K_{x_0}\right)\right] = \sum_{k=0}^{\infty} p_\text{c}\left(k\right) p_K\left(k\right), \tag{5.8}$$

where $p_K\left(k\right)$ is the probability mass function of K_{x_0}.

As the performance analysis of the studied network is quite challenging, in the following we make a few key approximations:

1. We consider the interference-limited networks and ignore the additive noise in the theoretical analysis.
2. The numbers of users in different cells are independent.
3. The served users form a homogeneous PPP, i.e., Φ'_u is an independent thinning of Φ_u.
4. The numbers of requests received by different BSs are independent, i.e., K_x for different active BSs are independent random variables.

Both Approximations 1 and 2 are commonly adopted in the study of Poisson wireless networks and have been applied in the analysis in Sects. 4.1–4.3. With Approximation 2, Φ'_b is an independent thinning process of Φ_b with density $\lambda_\text{b} p_\text{a}$, where p_a is the BS active probability given in (4.1). Thus, we directly obtain the density of $\Phi_\text{b}^{(1)}$ as

$$\lambda_\text{b}^{(1)} \approx \lambda_\text{b} p_\text{a} \mathbb{1}\left(\|x\| > \mu r_0\right), \tag{5.9}$$

where $\mathbb{1}\left(\|x\| > \mu r_0\right)$ is the indicator function that equals 1 if $\|x\| > \mu r_0$ and 0 otherwise. In this study, we not only need to know the distribution of active BSs but also the distribution of the served users, since it will determine the distribution of the number of requests received by an active BS. Thus, we made Approximation 3, which is analogous to Approximation 2. It was shown in [18] that the probability that a user is chosen to be served is $\frac{\lambda_b}{\lambda_u} p_a$, and thus, the density of Φ'_u is $p_a\lambda_b$. Finally, to obtain the distribution of $\Phi_b^{(2)}$, we adopt Approximation 4. Let ε denote the probability that the BS has received the request from the user but is unable to null interference for this user. Then, based on Approximation 4, the density of $\Phi_b^{(2)}$ is written as

$$\lambda_b^{(2)} \approx \varepsilon \lambda_b p_a \mathbb{1}\left(\|x\| \in [r_0, \mu r_0]\right). \tag{5.10}$$

Approximations 3 and 4 are introduced to decouple the distributions of active BSs and served users. This is mainly due to the intractability of considering the interaction between the BS process and the user process, which is needed to consider the user-centric interference nulling approach. With the help of *independent thinning* approximations, i.e., Approximations 2–4, accurate analytical results can be obtained for a transmission strategy involving close interaction between the BS and user processes.

With the above approximations, the receive SINR of the typical user is simplified as

$$\mathrm{SINR} = \frac{g_0 r_0^{-\alpha}}{\sum_{x\in\Phi_b^{(1)}\cup\Phi_b^{(2)}} g_x \|x\|^{-\alpha}}. \tag{5.11}$$

Next, we analyze the coverage probability based on (5.11). The accuracy of the approximations will be tested via simulations. For convenience, the key notations and symbols used in this section are listed in Table 5.1.

Table 5.1 Key notations and symbols used in Sect. 5.1. © 2015 IEEE. Reprinted, with permission, from [16]

Notation	Definition/Explanation
μ	IN range coefficient
K_x	# of requests received by the BS at x
b	# of feedback bits for one channel vector
ε	Probability that a BS receives a request from a user, but is unable to mitigate interference for this user
$p_c(k)$	The coverage probability to the user, whose home BS receives k requests
$p_{send}(y)$	Probability that the user at y sends a request to the BS at o
$p_K(k)$	The probability mass function of K_x
Φ_b, Φ_u	Set of BSs, and set of users
Φ'_b, Φ'_u	Set of active BSs and served users
$\Phi_b^{(1)}$, $\lambda_b^{(1)}(x)$	Set/density of interfering BSs farther than IN range
$\Phi_b^{(2)}$, $\lambda_b^{(2)}(x)$	Set/density of BSs who receive the request from the typical user, but are unable to do IN to this user
$\Phi_b^{(3)}$, $\lambda_b^{(3)}(x)$	Set/density of BSs who do IN to the typical user
$\Phi_u^{(1)}$, $\lambda_u^{(1)}(y)$	Set/density of users sending requests to the BS at o

5.1.3 Coverage Analysis: The Perfect CSI Case

In this subsection, we provide a tractable expression of the coverage probability. We assume that the home BS and the interfering BSs that need to suppress interference to the user have perfect CSI, while the effect of imperfect CSI will be investigated in the next subsection. It is shown from (5.8) that the coverage probability is composed of $p_\mathrm{c}(k)$ and the distribution of K_x, i.e., $p_K(k)$. Thus, we first derive $p_\mathrm{c}(k)$ and $p_K(k)$, which then lead an approximate expression of the coverage probability.

The Expression of $p_\mathrm{c}(k)$

From the SINR expression in (5.11), the coverage probability of the user whose home BS receives k request, i.e., $p_\mathrm{c}(k)$, is given by the following theorem.

Theorem 5.1 *The SIR coverage probability of the MISO small cell network is given by*

$$p_\mathrm{c}(k) = \left\| [(1 - p_\mathrm{a})\mathbf{I}_M + p_\mathrm{a}\mathbf{C}_M]^{-1} \right\|_1, \tag{5.12}$$

where $M = \max(N_\mathrm{t} - k, 1)$, *and the nonzero entries in the lower triangular Toeplitz matrix* $\mathbf{C}_M$ *as*

$$\begin{aligned} c_n &= \varepsilon(1-\mu^2)\mathbb{1}(n=0) + \varepsilon\frac{\delta\tau^n}{\delta - n}{}_2F_1\left(n+1, n-\delta; n+1-\delta; -\tau\right) \\ &+(1-\varepsilon)\frac{\delta\tau^n\mu^{2-n\alpha}}{\delta-n}{}_2F_1\left(n+1, n-\delta; n+1-\delta; -\tau\mu^{-\alpha}\right), \end{aligned} \tag{5.13}$$

for $0 \le n \le M-1$.

Proof According to the SINR expression (5.11), the log-Laplace transform consists of two parts, i.e., $\eta(s) = \eta_1(s) + \eta_2(s)$, where $\eta_1(s)$ and $\eta_2(s)$ correspond to the interference from $\Phi_\mathrm{b}^{(1)}$ and $\Phi_\mathrm{b}^{(2)}$, respectively.

Here, we follow the steps from (3.89)–(3.97) to prove the theorem. By incorporating the BS activity probability p_a and the IN range coefficient μ into consideration, the first part $\eta_1(s)$ is given by

$$\begin{aligned} \eta_1(s) &= -2\pi p_\mathrm{a}\lambda_\mathrm{b}\int_{\mu r_0}^{\infty}\left(1 - \mathbb{E}_g[\exp(-sgv^{-\alpha})]\right)v\mathrm{d}v \\ &= \pi p_\mathrm{a}\lambda_\mathrm{b}\mu^2 r_0^2 + \pi p_\mathrm{a}\lambda_\mathrm{b}\delta s^\delta \mathbb{E}_g\left[g^\delta\gamma(-\delta, s\mu^{-\alpha}r_0^{-\alpha}g)\right] \\ &= \pi p_\mathrm{a}\lambda_\mathrm{b}\mu^2 r_0^2 - \pi p_\mathrm{a}\lambda_\mathrm{b}\mu^2 r_0^2 \mathbb{E}_g\left[{}_1F_1\left(-\delta; 1-\delta; -s\mu^{-\alpha}r_0^{-\alpha}g\right)\right]. \end{aligned} \tag{5.14}$$

Then, the coefficients $\{t_{1,n}\}_{n=0}^{M-1}$ that are related to the n-th derivatives of the first part of the log-Laplace transform $\eta_1(s)$ are determined by

$$
\begin{aligned}
t_{1,n} &= \frac{(-s)^n}{n!}\eta_1^{(n)}(s) \\
&= -\pi p_{\mathrm{a}}\lambda_{\mathrm{b}}\mu^2 r_0^2 \frac{\delta}{\delta-n}\frac{\left(\tau\mu^{-\alpha}\right)^n}{n!} \qquad (5.15)\\
&\times\left\{\mathbb{E}_g\left[g^n{}_1F_1\left(n-\delta;n+1-\delta;-\frac{\tau\mu^{-\alpha}}{\theta}g\right)\right]-\mathbb{1}(n=0)\right\}\\
&= -\pi p_{\mathrm{a}}\lambda_{\mathrm{b}} r_0^2\left[c_{1,n}-\mathbb{1}(n=0)\right], \qquad (5.16)
\end{aligned}
$$

where

$$
\begin{aligned}
c_{1,n} &= \frac{\delta\mu^2}{\delta-n}\frac{\left(\tau\mu^{-\alpha}/\theta\right)^n}{n!}\mathbb{E}_g\left[g^n{}_1F_1\left(n-\delta;n+1-\delta;-\frac{\tau\mu^{-\alpha}}{\theta}g\right)\right],\\
&= \frac{\delta\tau^n\mu^{2-n\alpha}}{\delta-n}{}_2F_1\left(n+1,n-\delta;n+1-\delta;-\tau\mu^{-\alpha}\right), \qquad (5.17)
\end{aligned}
$$

where (5.17) applies Corollary 3.1 with $\theta=\kappa=\beta=1$. On the other hand, the second part of the log-Laplace transform $\eta_2(s)$ can be written as

$$
\begin{aligned}
\eta_2(s) &= -2\varepsilon\pi p_{\mathrm{a}}\lambda_{\mathrm{b}}\int_{r_0}^{\mu r_0}\left(1-\mathbb{E}_g[\exp(-sgv^{-\alpha})]\right)v\mathrm{d}v\\
&= -2\varepsilon\pi p_{\mathrm{a}}\lambda_{\mathrm{b}}\left(\int_{r_0}^{\infty}\left(1-\mathbb{E}_g[\exp(-sgv^{-\alpha})]\right)v\mathrm{d}v\right. \qquad (5.18)\\
&\quad\left.-\int_{\mu r_0}^{\infty}\left(1-\mathbb{E}_g[\exp(-sgv^{-\alpha})]\right)v\mathrm{d}v\right),
\end{aligned}
$$

which is the subtraction of two log-Laplace transforms that have similar forms as (5.14). Therefore, the coefficients $\{c_{2,n}\}_{n=0}^{M-1}$ of the second part of the log-Laplace transform $\eta_2(s)$ are determined by

$$
\begin{aligned}
c_{2,n} &= \varepsilon(1-\mu^2)\mathbb{1}(n=0)+\varepsilon\frac{\delta\tau^n}{\delta-n}{}_2F_1\left(n+1,n-\delta;n+1-\delta;-\tau\right) \qquad (5.19)\\
&\quad-\varepsilon\frac{\delta\tau^n\mu^{2-n\alpha}}{\delta-n}{}_2F_1\left(n+1,n-\delta;n+1-\delta;-\tau\mu^{-\alpha}\right).
\end{aligned}
$$

Therefore, the power series $T(z)$ in (3.94) is given by

$$
T(z) = \pi p_{\mathrm{a}}\lambda_{\mathrm{b}} r_0^2\left[1-C(z)\right], \qquad (5.20)
$$

where the coefficients $\{c_n\}_{n=0}^{M-1}$ of the power series $C(z)$ are given in (5.13) by $c_n = c_{1,n}+c_{2,n}$. With (5.20), the remaining proof is the same as that of Theorem 4.1.

The Expression of $p_K(k)$

To find the distribution of the number of requests received by a BS, we consider the typical BS located at the origin o. Then, this BS will receive a request from the user located at y if and only if $\|y\| \in (R, \mu R]$, where R is the distance from this user to its home BS and the pdf of R is given in (2.58) as

$$f_R(r) = 2\pi \lambda_b r e^{-\pi \lambda_b r^2}. \tag{5.21}$$

Therefore, the probability that a served user at y will send a request to the BS at o can be derived as

$$\begin{aligned} p_{\text{send}}(y) &= \mathbb{P}(R < \|y\| \le \mu R) = \int_{\frac{\|y\|}{\mu}}^{\|y\|} f_R(r)\,\mathrm{d}r \\ &= e^{-\pi \lambda_b \frac{\|y\|^2}{\mu^2}} - e^{-\pi \lambda_b \|y\|^2}. \end{aligned} \tag{5.22}$$

Based on Approximation 3, we obtain that the set of users who send the request to the BS at o, denoted by $\Phi_u^{(1)}$, is a location-dependent thinned PPP of Φ_u' with density $p_a \lambda_b p_{\text{send}}(y)$. Therefore, the number of requests received by the BS at o is the total number of the users in $\Phi_u^{(1)}$. The distribution of the number of requests received by this BS is given in Lemma 5.1.

Lemma 5.1 *With Approximation 3, the number of requests received by a BS, i.e., $p_K(k)$, is a Poisson distribution with mean $\bar{K} = p_a(\mu^2 - 1)$, i.e.,*

$$p_K(k) = \frac{(\bar{K})^k}{k!} e^{-\bar{K}}. \tag{5.23}$$

Proof See Appendix.

Remark 5.2 (*Accuracy of Approximation 3*) The main source of the inaccuracy of the numerical results is the Poisson assumption for K_x, i.e., the number of requests received by a BS, which is derived based on Approximation 3, assuming independent thinning of the served users. The comparisons of the simulation results and the Poisson approximations are shown in Figs. 5.2, 5.3 and 5.4. We observe that the approximation is less accurate with large μ and small ρ.

The Expression of p_c

Finally, based on $p_K(k)$, we derive the expression of ε, i.e., the probability that the BS has received the request from the typical user but is unable to null interference for the typical user. We consider an interfering BS from the annulus $[r_0, \mu r_0]$ chosen uniformly at random. Besides the request from the typical user, we assume it

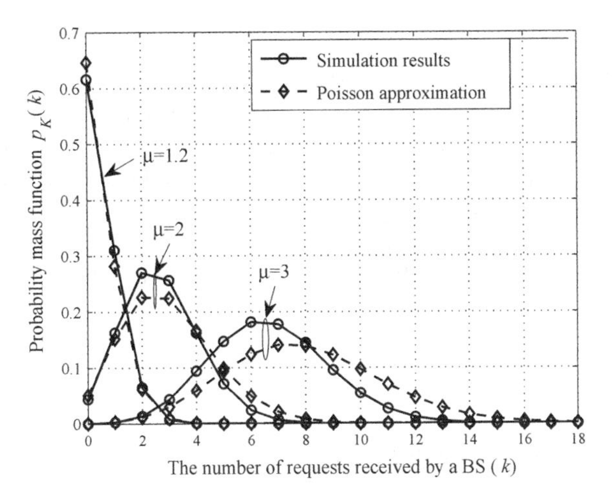

Fig. 5.2 The probability mass function of K_x for $\rho = 0.1$ ($p_a \approx 0.99$). © 2015 IEEE. Reprinted, with permission, from [16]

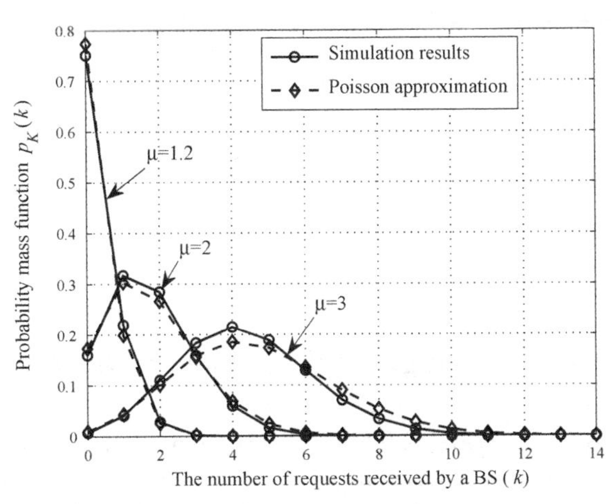

Fig. 5.3 The probability mass function of K_x for $\rho = 1$ ($p_a \approx 0.58$). © 2015 IEEE. Reprinted, with permission, from [16]

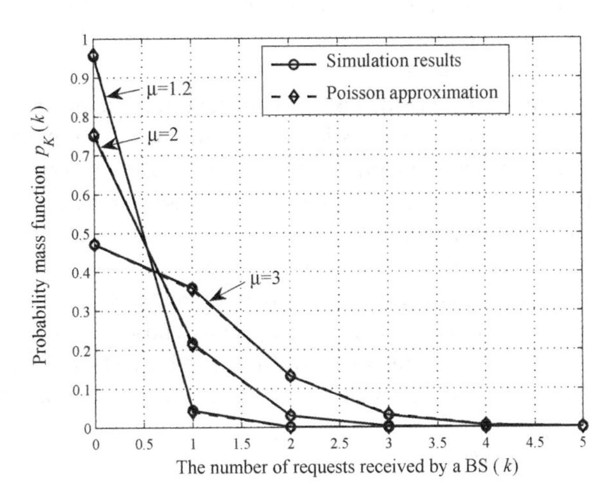

Fig. 5.4 The probability mass function of K_x for $\rho = 10$ ($p_a \approx 0.09$). © 2015 IEEE. Reprinted, with permission, from [16]

receives $\hat{K}$ more requests from other users. If $\hat{K} = \hat{k} \geq N_\mathrm{t} - 1$, then with probability $\frac{\hat{k}+1-(N_\mathrm{t}-1)}{\hat{k}+1}$ this BS will not perform interference nulling for the typical user, as in this case, the BS will randomly pick $N_\mathrm{t} - 1$ from $\hat{k} + 1$ requests for interference nulling. From (5.22), we show that each served user will send the request based on its own distance to the interfering BS, which is independent of the other served users. Thus, given that the typical user has sent the request to the interfering BS, the number of other requests received by this BS $\hat{K}$ still follows the Poisson distribution with the probability mass function (5.23). Therefore, the expression of ε is given by

$$\varepsilon = \sum_{k=N_\mathrm{t}-1}^{\infty} \frac{k+1-(N_\mathrm{t}-1)}{k+1} p_K(k). \tag{5.24}$$

By substituting (5.12) and (5.23) into (5.8), the coverage probability to the typical user is given by

$$\begin{aligned} p_\mathrm{c} \approx & \left\| \left[(1-p_\mathrm{a})\mathbf{I}_M + p_\mathrm{a}\mathbf{C}_M\right]^{-1} \right\|_1 \frac{\left[p_\mathrm{a}\left(\mu^2-1\right)\right]^k}{k!} e^{-p_\mathrm{a}\left(\mu^2-1\right)} \\ & + \frac{\gamma\left(N_\mathrm{t}, p_\mathrm{a}\left(\mu^2-1\right)\right)}{(N_\mathrm{t}-1)!\,(1+p_\mathrm{a}c_0)}, \end{aligned} \tag{5.25}$$

where $\gamma(s, x)$ is the lower incomplete gamma function.

5.1.4 Coverage Analysis: The Limited Feedback Case

The coverage probability in (5.25) was derived assuming perfect CSI. However, there will always be inaccuracy in the available CSI, which will degrade the performance. In this subsection, we consider the case where the active BSs will obtain quantized CSI through limited feedback, which is a common technique to provide CSI at the transmitter side [19].

With limited feedback, the channel direction information (CDI) is fed back using a quantization codebook known at both the transmitter and receiver [19]. The quantization is chosen from a codebook of unit norm vectors of size 2^b, where b is the number of feedback bits for each channel. We assume that each user uses a different codebook to avoid getting the same quantization vector for different channels. The codebook for the typical user is denoted as $\mathscr{C}_o = \left\{\bar{\mathbf{c}}_j : j = 1, 2, \ldots, 2^b\right\}$, where the codewords are generated using random vector quantization (RVQ), i.e., each quantization vector $\bar{\mathbf{c}}_j$ is independently chosen from the isotropic distribution on the N_t-dimensional unit sphere [20]. It has been shown in [20] that RVQ can facilitate the analysis and provide performance close to the optimal quantization. Each user quantizes its CDI to the closest codeword, measured by the inner product. Therefore, the quantized CDI is

$$\hat{\mathbf{h}}_x = \arg\max_{\bar{\mathbf{c}}_j \in \mathscr{C}_o} \left|\bar{\mathbf{h}}_x^H \bar{\mathbf{c}}_j\right|, \tag{5.26}$$

where $\bar{\mathbf{h}}_x \triangleq \frac{\mathbf{h}_x}{\|\mathbf{h}_x\|}$ is the actual CDI. Then, the index of the quantized CDI $\hat{\mathbf{h}}_x$ is fed back with b bits. In this subsection, we assume the feedback channel is error-free and without delay. Thus, each active BS will use the quantized CDI of both the signal and interference channels to design its transmission vector.

For user-centric intercell interference nulling, each user not only needs to feed back CDI to its home BS but also to the coordinating BSs. We assume that each user feeds back all the quantized CDI to its home BS, and then the home BS forwards the associated CDI to the corresponding BSs through backhaul connection. With the imperfect CSI at each BS, the received signal for the typical user (at the origin o) is given by

$$\hat{s}_{x_0} = \sqrt{P_t r_0^{-\frac{\alpha}{2}}}\mathbf{h}_{x_0x_0}\hat{\mathbf{f}}_{x_0}s_{x_0} + \sum_{x\in\Phi'=\Phi_b^{(1)}\cup\Phi_b^{(2)}\cup\Phi_b^{(3)}} \sqrt{P_t}\,\|x\|^{-\frac{\alpha}{2}}\,\mathbf{h}_{x_0x}\hat{\mathbf{f}}_x s_x + n_{x_0}, \tag{5.27}$$

where $\Phi_b^{(3)}$ denotes the set of interfering BSs who help this user to suppress interference. It can be shown that $\Phi_b^{(3)}$ is a non-homogeneous PPP with density $\lambda_3(x)$ given by

$$\lambda_3(x) = (1-\varepsilon)\,p_a\lambda_b\mathbb{1}\left(\|x\| \in [r_0, \mu r_0]\right). \tag{5.28}$$

Note that in the perfect CSI case, the BSs in $\Phi_b^{(3)}$ do not cause interference to the user. However, with limited feedback, there is residual interference from these BSs due to the quantization error. Moreover, the precoding vector of the BS at x, denoted as $\hat{\mathbf{f}}_x$, has the same expression as (5.4) but is designed based on quantized CDI. Therefore, the receive SINR can be written as

$$\mathrm{SIR} = \frac{\hat{g}_0 r_0^{-\alpha}}{\sum_{x\in\Phi_b^{(1)}\cup\Phi_b^{(2)}\cup\Phi_b^{(3)}} \hat{g}_x\,\|x\|^{-\alpha}}, \tag{5.29}$$

where the equivalent channel gain is given as $\hat{g}_x = \left|\mathbf{h}_x^H\hat{\mathbf{f}}_x\right|^2$. Compared with the SINR expression (5.11) in the perfect CSI case, it is clear that due to limited feedback, there is another part of interference, which is $\sum_{x\in\Phi_b^{(3)}} \hat{g}_x\,\|x\|^{-\alpha}$. Moreover, $\hat{g}_0$ no longer follows the gamma distribution with scale parameter 1, and its parameter will depend on the number of feedback bits b.

To determine the coverage probability, we first need to get the distribution of channel gains with limited feedback. Since the precoding vector $\hat{\mathbf{f}}_x$ is independent with the channel from the BS at $x \in \Phi_b^{(1)} \cup \Phi_b^{(2)}$, $\hat{g}_x$ is still exponential, i.e., $\hat{g}_x \sim \mathrm{Exp}(1)$ for $x \in \Phi_b^{(1)} \cup \Phi_b^{(2)}$. However, due to the quantization error, the distributions of the information channel gain $\hat{g}_0$ and the residual interference channel gain $\hat{g}_x$ for $x \in \Phi_{\mathrm{b}}^{(3)}$ will change, which are given in the following lemma.

Lemma 5.2 *Given the number of feedback bits for one channel vector as b, the distribution of the information channel gain $\hat{g}_0$ can be approximated as*

$$\hat{g}_0 \sim \mathrm{Gamma}\left(M, \kappa_0\right), \tag{5.30}$$

where $M = \max\left(N_{\mathrm{t}} - k, 1\right)$, k is the number of requests received by this BS, $\kappa_0 \triangleq 1 - 2^b B\left(2^b, \frac{N_{\mathrm{t}}}{N_{\mathrm{t}}-1}\right)$, and $B\left(x, y\right) = \frac{\Gamma(x)\Gamma(y)}{\Gamma(x+y)}$ is the beta function.

Moreover, the residual interference channel gain $\hat{g}_x$ for $x \in \Phi_{\mathrm{b}}^{(3)}$ can be approximated as

$$\hat{g}_x \sim \mathrm{Exp}\left(1/\kappa_I\right) = \mathrm{Gamma}(1, \kappa_I), \tag{5.31}$$

where $\kappa_I = 2^b B\left(2^b, \frac{N_{\mathrm{t}}}{N_{\mathrm{t}}-1}\right) = 1 - \kappa_0$ is the quantization distortion.

Proof The proof can be found in [8, Lemma 5].

Then, following the same procedure of Theorem 5.1 and using the distributions of $\hat{g}_0$ and $\hat{g}_x$ in Lemma 5.2, the coverage probability to the typical user is given in the following proposition.

Proposition 5.1 *With limited feedback, the coverage probability to the user whose home BS receives k requests is given by*

$$p_{\mathrm{c,LF}}\left(k\right) = \left\| \left[(1 - p_{\mathrm{a}})\mathbf{I}_M + p_{\mathrm{a}}\mathbf{C}_M\right]^{-1} \right\|_1, \tag{5.32}$$

and the nonzero entries in the lower triangular Toeplitz matrix $\mathbf{C}_M$ as

$$\begin{aligned} c_{n,\mathrm{lF}} = {} & (1 - \mu^2)\mathbb{1}(n = 0) + \varepsilon\frac{\delta\tau^n}{\delta - n}{}_2F_1\left(n+1, n-\delta; n+1-\delta; -\tau\right) \\ & +(1-\varepsilon)\frac{\delta\tau^n\mu^{2-n\alpha}}{\delta - n}{}_2F_1\left(n+1, n-\delta; n+1-\delta; -\tau\mu^{-\alpha}\right) \\ & +(1-\varepsilon)\frac{\delta}{\delta - n}\left(\frac{\tau\kappa_I}{\kappa_0}\right)^n {}_2F_1\left(n+1, n-\delta; n+1-\delta; -\frac{\tau\kappa_I}{\kappa_0}\right) \\ & -(1-\varepsilon)\frac{\delta\mu^{2-n\alpha}}{\delta - n}\left(\frac{\tau\kappa_I}{\kappa_0}\right)^n {}_2F_1\left(n+1, n-\delta; n+1-\delta; -\frac{\tau\kappa_I}{\kappa_0}\mu^{-\alpha}\right), \end{aligned} \tag{5.33}$$

for $0 \le n \le M - 1$.

Proof Based on the SIR expression in (5.29), the log-Laplace transform consists of three parts, i.e.,

$$\eta\left(s\right) = \eta_1\left(s\right) + \eta_2\left(s\right) + \eta_3\left(s\right), \tag{5.34}$$

where $\eta_i(s)$ corresponds to the interference from $\Phi_{\mathrm{b}}^{(i)}$ for $i = 1, 2, 3$. Note that $\eta_1(s)$ is the same as (5.14), while $\eta_2(s)$ is the same as (5.18), since the interference distribution from $\Phi_{\mathrm{b}}^{(1)}$ and $\Phi_{\mathrm{b}}^{(2)}$ are the same as the case with the perfect CSI. The third part $\eta_3(s)$ is given by

$$\begin{aligned}\eta_3(s) &= -2\pi(1-\varepsilon)\, p_{\mathrm{a}}\lambda_{\mathrm{b}} \int_{r_0}^{\mu r_0} \left(1 - \mathbb{E}_{\hat{g}}[\exp(-s\hat{g}v^{-\alpha})]\right) v\mathrm{d}v \\ &= -2\pi(1-\varepsilon)\, p_{\mathrm{a}}\lambda_{\mathrm{b}} \Bigg(\int_{r_0}^{\infty} \left(1 - \mathbb{E}_{\hat{g}}[\exp(-s\hat{g}v^{-\alpha})]\right) v\mathrm{d}v \\ &\quad - \int_{\mu r_0}^{\infty} \left(1 - \mathbb{E}_{\hat{g}}[\exp(-s\hat{g}v^{-\alpha})]\right) v\mathrm{d}v \Bigg).\end{aligned} \tag{5.35}$$

Follow the derivation in (5.18) and Corollary 3.1, we obtain the coefficients $\{c_{3,n}\}_{n=0}^{M-1}$ of the third part of the log-Laplace transform $\eta_3(s)$, given by

$$\begin{aligned}c_{3,n} &= (1-\varepsilon)(1-\mu^2)\mathbb{1}(n=0) + (1-\varepsilon)\frac{\delta}{\delta-n}\left(\frac{\tau\kappa_I}{\kappa_0}\right)^n {}_2F_1\left(n+1, n-\delta; n+1-\delta; -\frac{\tau\kappa_I}{\kappa_0}\right) \\ &\quad -(1-\varepsilon)\frac{\delta\mu^{2-n\alpha}}{\delta-n}\left(\frac{\tau\kappa_I}{\kappa_0}\right)^n {}_2F_1\left(n+1, n-\delta; n+1-\delta; -\frac{\tau\kappa_I}{\kappa_0}\mu^{-\alpha}\right).\end{aligned} \tag{5.36}$$

By substituting (5.32) into (5.8), we obtain the final expression of the coverage probability with limited feedback.

Remark 5.3 By comparing the expressions of the coverage probability of the perfect CSI case and the limited feedback case, we observe that the only terms changed are $\{c_n\}_{n=0}^{M-1}$. Moreover, the quantization distortion κ_I decreases when increasing the number of feedback bits b, and $\kappa_I \to 0$ when $b \to \infty$. This means that $\{c_{n,\mathrm{LF}}\}_{n=0}^{M-1}$ in (5.33) will converge to $\{c_n\}_{n=0}^{M-1}$ in (5.13) for the perfect CSI case as b increases.

5.1.5 Optimization: To Determine the Interference Nulling Range

By now, we have obtained an approximation of the coverage probability using the user-centric intercell interference nulling strategy with a fixed μ. We then search for the optimal μ numerically, which is the μ that maximizes the coverage probability, i.e.,

$$\mu^\star = \arg\max_{\mu} p_{\mathrm{c}}. \tag{5.37}$$

In this subsection, we use $\mu^\star$ and $\hat{\mu}^\star$ to denote the optimal values obtained through simulation and based on the approximation in (5.25), respectively. A main benefit of the presented analytical approach is that $\hat{\mu}^{\star}$ can be found much more efficiently than $\mu^{\star}$, which requires extensive simulations. Therefore, based on numerical results, we can examine the effectiveness of this strategy and observe the effects of different parameters, such as the BS density and the number of antennas, on the network performance.

Remark 5.4 (*The effect of the BS and user densities*) It is apparent from (5.25) that the effect of λ_{b} and λ_{u} on p_{c} is determined by the BS-user density ratio ρ. In the following of this subsection, we will vary ρ to investigate the effect of the BS density or the user density. Increasing ρ can be viewed as increasing the BS density for a given user density, or equivalently, as decreasing the user density with a certain BS density.

Remark 5.5 (*The non-coordination strategy*) Note that when $\mu = 1$, (5.25) becomes an exact expression, i.e., $p_{\mathrm{c}} = \left\| [(1 - p_{\mathrm{a}})\mathbf{I}_M + p_{\mathrm{a}}\mathbf{C}_M]^{-1} \right\|_1$, where $\{c_n\}_{n=0}^{M-1}$ are given in (5.13) with $\mu = 1$ and $\varepsilon = 0$. The result for this special case was the same as (4.6). In this subsection, we refer the performance of the *non-coordination strategy* as p_{c} in (5.25) for $\mu = 1$.

In Fig. 5.5, we compare the simulation results with the approximation results, where the BS density is $\lambda_{\mathrm{b}} = 0.001$ per m^2, user density is $\lambda_{\mathrm{u}} = 0.01$ per m^2 and the SINR threshold is $\tau = 10$. From Fig. 5.5, we infer that selecting a proper IN range coefficient μ can greatly improve the network performance and that there exists an optimal μ to achieve the maximum coverage probability. Particularly, compared with the non-coordination scenario (i.e., $\mu = 1$), using the user-centric intercell interference nulling with the optimal μ can improve the relative performance by about 37%, which indicates the effectiveness of the presented interference nulling strategy. Moreover, by comparing the simulation results with the approximation, we find that the approximation result is lower than the simulation, and the approximation error increases with μ. This is because the approximated $p_K(k)$ is less accurate when μ

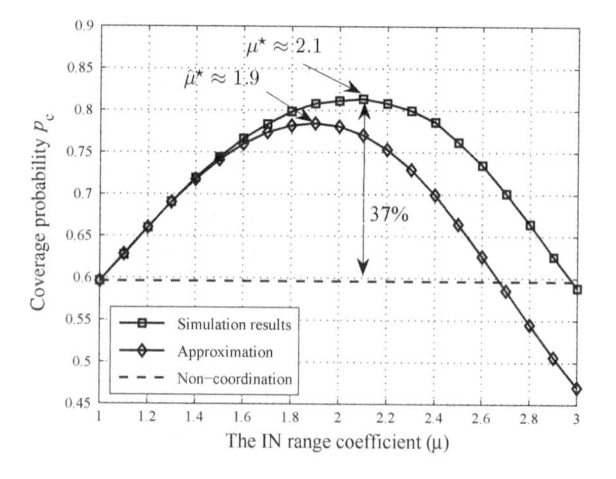

Fig. 5.5 The coverage probability as a function of μ, with $\lambda_{\mathrm{b}} = 10^{-3}\mathrm{m}^{-2}$, $\lambda_{\mathrm{u}} = 10^{-2}\mathrm{m}^{-2}$, $N_{\mathrm{t}} = 8$, $\alpha = 4$, and $\tau = 10$. The maximum performance gain of 37% is the relative improvement from 60 to 82%. ©2015 IEEE. Reprinted, with permission, from [16]

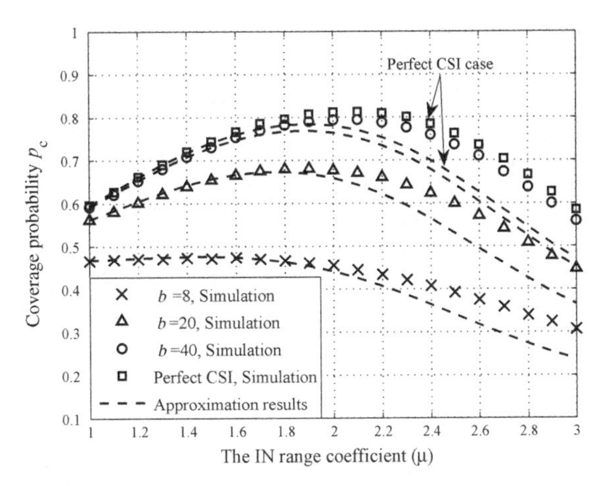

Fig. 5.6 The coverage probability as a function of μ, with $\lambda_b = 10^{-3}\text{m}^{-2}$, $\lambda_u = 10^{-2}\text{m}^{-2}$, $N_t = 8$, $\alpha = 4$, and $\tau = 10$. The dashed lines are the approximation results for $b = 8, 20, 40$, and the perfect CSI case, respectively. ©2015 IEEE. Reprinted, with permission, from [16]

is large. However, it is also shown that the optimal IN range coefficient μ obtained from the approximation ($\hat{\mu}^\star \approx 1.9$) is close to the optimal value from simulations ($\mu^\star \approx 2.1$). As the curve of p_s is quite flat near $\mu^\star$, a small deviation of $\hat{\mu}^\star$ will only slightly affect p_s, and thus, we obtain a near-optimal μ via the approximate expression.

In Fig. 5.6, we show the effect of the number of feedback bits b on the coverage probability. We see that with limited feedback, the coverage probability is still a quasi-concave function with respect to the IN range coefficient μ, i.e., p_c will first increase and then decrease when μ increases. Moreover, similar to the perfect CSI case, the approximation is more accurate when μ is small, but for different values of b, the optimal μ obtained via the approximation is close to the one via simulation. Thus, the approximate result can help to optimize the presented interference nulling strategy.

Next, we will compare the presented strategy with other interference nulling methods, and then present some numerical results to provide guidelines for practical system design. First, we compare the presented interference nulling strategy with other interference nulling methods. One method for comparison is the fixed-number-based IN. Specifically, each user requests N nearest interfering BSs to suppress interference. But if the BS receives more than $N_t - 1$ requests, it will randomly choose $N_t - 1$ users to mitigate interference. The other method for comparison is the random BS clustering method, proposed in [10].

Figure 5.7 shows the comparison of results for different BS-user density ratios. Note that for all the methods, we use the optimal value of the key parameter, i.e., for the presented interference nulling strategy, we use the optimal IN range coefficient μ. For the fixed-number-based IN, we optimize N to obtain the maximum

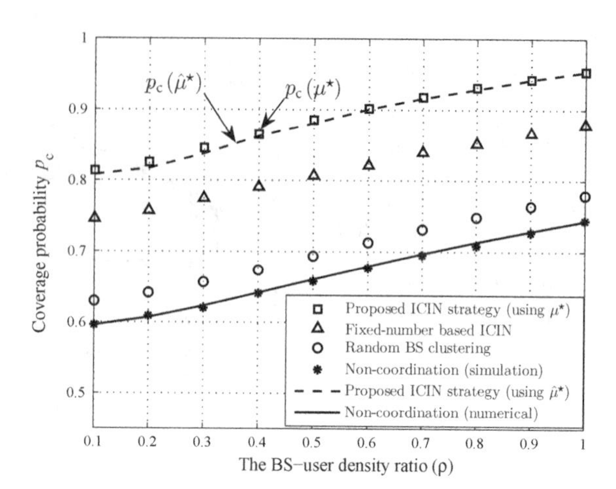

Fig. 5.7 The coverage probability with different BS-user density ratios, with $\lambda_b = 10^{-3}\text{m}^{-2}$, $N_t = 8$, $\alpha = 4$, and $\tau = 10$, where "ICIN" stands for intercell interference nulling. The dashed line is obtained through simulation using the approximated optimal μ (i.e., $\hat{\mu}^{\star}$) from (5.25), while the square is obtained by using the optimal μ (i.e., $\mu^{\star}$) in simulation. ©2015 IEEE. Reprinted, with permission, from [16]

p_c. And for random BS clustering, we find the optimal cluster size.[2] Moreover, the coverage probability without coordination is presented as the baseline. From Fig. 5.7, we can find that: (1) User-centric coordination methods significantly outperform the BS clustering method, and the presented strategy performs better than the fixed-number-based IN. (2) Using the approximated optimal μ (denoted as $\hat{\mu}^{\star}$) provides performance very close to that using simulation to search the optimal μ (denoted as $\mu^{\star}$), so it can be used in practice. (3) As ρ increases,[3] the coverage probability increases, and it appears that the performance gaps between different methods stay constant.

The superior performance of the presented interference nulling strategy is because it can more effectively identify the dominant interference for each user, while the other two coordination methods are not adaptive to each user's interference situation. Furthermore, we find that the optimal μ in Fig. 5.8 is about 2, from which we derive the average number of requests $\bar{K}$ to be around 2–3, so most BSs can well handle the requests with the available spatial degrees of freedom. This also confirms that the presented strategy is practical since the IN range is not large, and thus, the amount of signaling overhead will be acceptable.

Next, we shall provide some design guidelines for the practical network deployment with the presented interference nulling strategy. We will consider two different options to improve the network performance, i.e., to deploy more BSs or to increase the number of BS antennas. The effects of these two approaches are shown in Fig. 5.9. In this figure, for a given value of N_t, we obtain the minimal BS-user density ratio ρ required to achieve the coverage probability of 0.9 with the SINR threshold as 0

---

[2]The optimal N of the number-based ICIN and the optimal cluster size for random BS clustering can only be obtained via simulation as no analytical expression of p_c is available for these two cases.

[3]Note that many previous works such as focused on the case $p_a = 1$, which could not capture the effect of the user distribution.

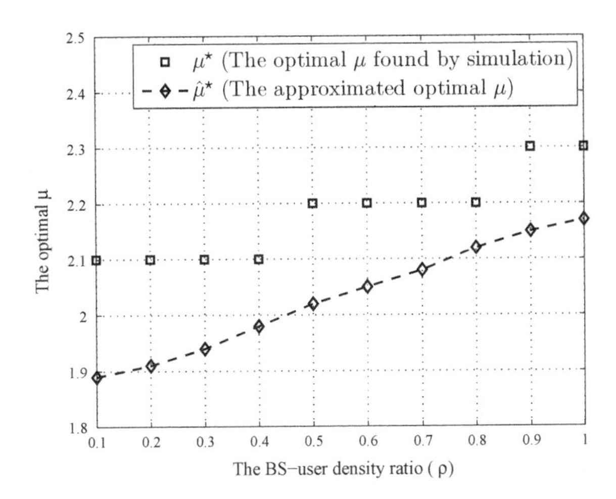

Fig. 5.8 The optimal μ obtained through simulation and approximation with different BS-user density ratio ρ, with $\lambda_b = 10^{-3}\text{m}^{-2}$, $N_t = 8$, $\alpha = 4$, and $\tau = 10$. ©2015 IEEE. Reprinted, with permission, from [16]

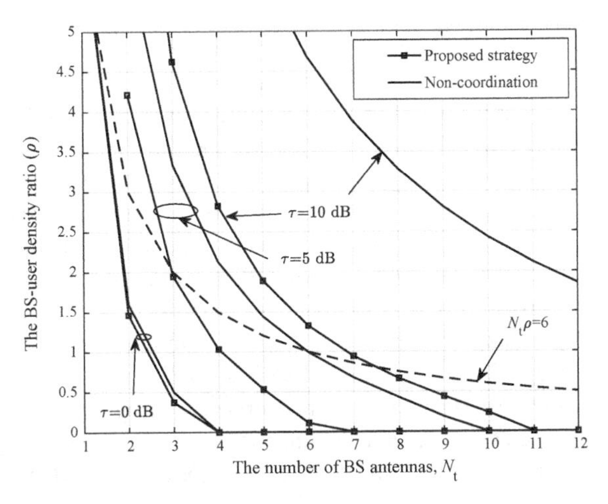

Fig. 5.9 The minimal BS-user density ratio required to achieve $p_c = 0.9$, with different numbers of BS antennas, with $\alpha = 4$. For the presented strategy, the approximated optimal μ (i.e., $\hat{\mu}^\star$) is used to obtain p_c, then we find the minimal ρ to achieve $p_c = 0.9$. The dashed line is the reference line, on which all the points have the same value of $N_t\rho$. ©2015 IEEE. Reprinted, with permission, from [16]

dB, 5 dB, and 10 dB, respectively. The approximated optimal μ (i.e., $\hat{\mu}^\star$) is used for the interference nulling strategy. The following interesting and insightful observations can be made: (1) When the SINR threshold τ is small, the performance of the presented strategy is similar to the performance of the non-coordination strategy. It can be found that the optimal μ tends to 1 when τ decreases. (2) When τ is large, the advantage of the presented interference nulling strategy is significant. For example, with $N_t = 6$ and $\tau = 10$ dB, it can achieve the same performance as the non-coordination strategy with only 1/3 of the BS density; while with the BS-user density ratio as 2, it achieves the same performance at $N_t = 5$ instead of $N_t = 12$. It means that the deployment cost can be greatly reduced with the presented interference nulling strategy to meet the requirement of high data rate transmission. (3) The number of BS antennas plays a more important role than the BS density. If we fix the total number of antennas per unit area, e.g., fix $\rho N_t = 6$ as in Fig. 5.9, it is

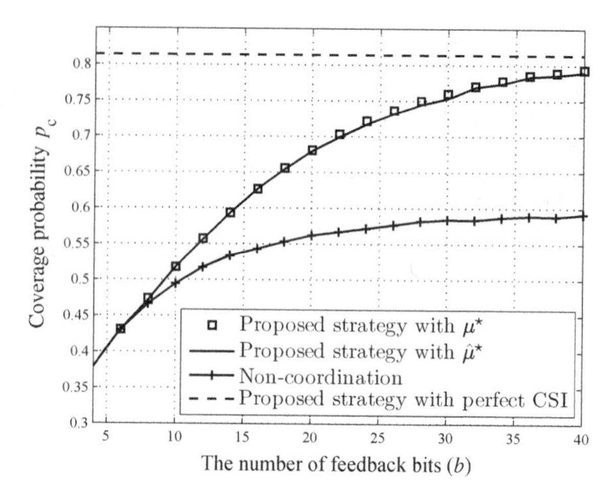

Fig. 5.10 The coverage probability with different b, with $\lambda_b = 10^{-3}\text{m}^{-2}$, $\lambda_u = 10^{-2}\text{m}^{-2}$, $N_t = 8$, $\alpha = 4$, and $\tau = 10$. ©2015 IEEE. Reprinted, with permission, from [16]

shown that increasing N_t can improve the supported SINR threshold, which implies that colocated BS antennas can support higher data rate requirement.[4]

So far we have demonstrated the effectiveness of the presented user-centric intercell interference nulling strategy, especially for the high SINR requirement (i.e., τ is high). Next, we will investigate the effect of the limited feedback on the performance. In Fig. 5.10, we evaluate the effect of the number of feedback bits b, where the performances of cooperative and noncooperative systems are compared with different values of b. We find that when the number of feedback bits b increases, the coverage probability will approach the perfect CSI case. However, if b is not sufficiently large, using interference nulling has a similar performance with the non-coordination strategy. This is because when b is small, the quantization error is large, which will limit the performance of interference nulling. Thus, a sufficient number of bits are required to quantize each channel vector in order to exploit the performance gains of interference nulling, e.g., $b \geq 10$ for the network considered in Fig. 5.10.

5.1.6 Summary

In this section, we presented an interference nulling strategy to demonstrate the effectiveness of the general analytical framework for designing interference management techniques. Without the tractable expression of the coverage probability, it would take much longer time to find the optimal IN range coefficient through simulations, and it would be almost impossible to find the design insights, such as the ones shown in Fig. 5.9. Furthermore, the analytical framework applies to both the perfect CSI case and the limited feedback case. From this study, it was shown that

[4]This conclusion depends on the actual transmission strategy, and a full comparison of colocated and distributed antenna deployment is left to future work.

the presented user-centric interference nulling strategy provides significant performance gains compared with the case without coordination, and outperforms other interference nulling methods. In addition, it still provides remarkable performance gains even with a moderate number of feedback bits. Finally, it demonstrated that to achieve high SINR requirement, it is indispensable to apply interference management in large-scale wireless networks.

5.2 ASE and Link Reliability Trade-Off in HetNets

The results presented in Chap. 4 and Sect. 5.1 considered homogeneous networks, where all the BSs are of the same type. In 5G networks, different types of BSs will be deployed and overlaid with the existing macrocell network. Such networks are called heterogeneous networks (HetNets), which can significantly improve the spatial reuse and thus the ASE, as well as providing uniform coverage [21–23]. In this section, we present analytical results for general multi-antenna HetNets and discuss a few new features. In particular, the analysis reveals a unique trade-off between the ASE and link reliability that does not exist in homogeneous networks or single-antenna HetNets. This insight is then applied to optimize the network performance.

5.2.1 A General Multi-Antenna HetNet Model

In this subsection, we present a general HetNet model, which can be used to analyze the network with arbitrary types of BSs. Consider a downlink cellular network consisting of K different tiers of BSs, indexed by $\mathcal{K} = \{1, 2, \ldots, K\}$. In the k-th tier, the BSs are spatially distributed according to a homogeneous Poisson point process (PPP) in $\mathbb{R}^2$ of density λ_k, denoted by Φ_k. Each BS in the k-th tier has transmit power P_k, with N_k antennas, and serves U_k ($U_k \leq N_k$) users at each time slot, i.e., intra-cell SDMA is considered. Note that when $U_k = 1, \forall k \in \mathcal{K}$, the network becomes a TDMA HetNet. Meanwhile, we assume that mobile users are distributed as a homogeneous PPP Φ_u of density λ_u, which is independent of $\Phi_k, \forall k \in \mathcal{K}$. Each user has a single receive antenna, and is associated with one BS.[5] We assume the network is fully loaded ($\lambda_\mathrm{u} \gg \lambda_k, \forall k \in \mathcal{K}$), i.e., there are at least U_k users in each cell in the k-th tier, which is a common assumption in the analysis of random cellular networks [24]. An example of a three-tier multiuser MIMO HetNet is shown in Fig. 5.11.

The network is open access, which means a user is allowed to access any BS in any tier. Particularly, each user will listen to the pilot signals from different BSs, and

[5]Note that the analysis also applies to single-user MIMO, where each user in the k-th tier has U_k uncorrelated receive antennas, and BSs in the k-th tier apply equal power allocation to the U_k streams. Then the coverage probability to the typical user analyzed in this section becomes the coverage probability per stream.

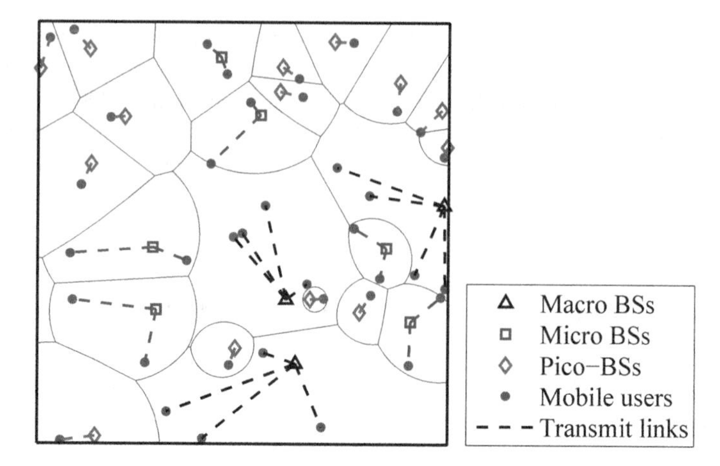

Fig. 5.11 A demonstration of a three-tier HetNet, in which each macro-BS has eight antennas and serves four users, each micro-BS has four antennas and serves two users, and each pico-BS has a single antenna and serves one user. ©2016 IEEE. Reprinted, with permission, from [25]

measure the long-term average received power. We consider a commonly adopted cell association rule [26], i.e., the user will be associated with the k-th tier if[6]

$$k = \arg\max_{j \in \mathcal{K}} P_j B_j r_j^{-\alpha}, \tag{5.38}$$

where r_j denotes the minimal distance from a user to its nearest BS in the j-th tier, $\alpha > 2$ is the path loss exponent, and B_j is the bias factor of the j-th tier, which is used for load balancing [26]. It is shown in [26] that the typical user will be associated with the k-th tier with probability

$$A_k = \frac{\lambda_k P_k^{\delta} B_k^{\delta}}{A}, \tag{5.39}$$

where $A = \sum_{j=1}^{K} \lambda_j P_j^{\delta} B_j^{\delta}$.

For each BS, equal power allocation is assumed, i.e., the BS in the k-th tier will equally allocate P_k to its U_k users. Furthermore, we adopt ZF precoding assuming perfect CSI at each BS. Therefore, the received power of the typical user at origin o from a BS located at $x \in \Phi_k$ is given by

$$P(x) = \frac{P_k}{U_k} g_{x,k} \|x\|^{-\alpha}, \tag{5.40}$$

[6]In the user association procedure, the first antenna normally uses the total transmission power of a BS to transmit reference signals for biased received power determination according to the LTE standard [27].

where $g_{x,k}$ is the channel gain, and its distribution depends on whether the BS is the home BS (i.e., the associated BS) or the interfering BS. Assuming Rayleigh fading channels, it is shown in Sect. 3.1 that $g_{x,k}$ is gamma distributed with the shape parameter $N_k - U_k + 1$ if the BS at x is the home BS, i.e., the distribution of the signal power gain in (3.16) is given by

$$\frac{P_k}{U_k} g_{x_0,k} \sim \text{Gamma}\left(M_k, \frac{P_k}{U_k}\right), \tag{5.41}$$

where $M_k \triangleq N_k - U_k + 1$, while the interferer's power gain in (3.1) can be approximated as

$$\frac{P_j}{U_j} g_{x,j} \sim \text{Gamma}\left(U_j, \frac{P_j}{U_j}\right) \tag{5.42}$$

if the BS is the interfering BS. This approximation has been commonly adopted in performance analysis of multiuser MIMO networks, e.g., [28–31]. Thus in multiuser MIMO HetNets, the number of users served by each BS in one tier will affect both the information signal quality to its own users and the interference caused to users in other cells.

We assume that universal frequency reuse is applied, which implies that the user will not only receive the information signal from its home BS but also suffer interference from all the other BSs. The resulting SIR of the typical user, served by the BS at x_0 in the k-th tier, is given by

$$\text{SIR}_k = \frac{\frac{P_k}{U_k} g_{x_0,k} r_k^{-\alpha}}{\sum_{j=1}^{K} \sum_{x \in \Phi'_j} \frac{P_j}{U_j} g_{x,j} \|x\|^{-\alpha}}. \tag{5.43}$$

Note that we only consider SIR in this section. The analysis can be extended to SINR at a slight loss of tractability. But the SIR distribution generally captures the key trade-offs and is realistic for reasonably dense deployments.

Remark 5.6 In summary, the most important parameters of the network are the per-tier parameters: $\{\lambda_k, P_k, B_k, N_k, U_k\}$. Compared with existing works, the K-tier HetNet model presented in this monograph includes intracell multi-antenna transmissions, which generalizes the SISO HetNet model (where $N_k = U_k = 1$ for all $k \in \mathcal{K}$) [26]. Based on this model, the effects of the multiple antennas and the number of served users can be evaluated. Meanwhile, both the signal and interfering channel gains become gamma distributed, while in SISO HetNets, all the channel gains are exponentially distributed.

As introduced in Sect. 2.2, the coverage probability and ASE are two different aspects of a communication system, and both of them are fundamental metrics. The coverage probability also stands for the success probability of the transmission link, which measures the link reliability. A higher success probability implies that the transmission links are more reliable, i.e., a better quality of experience (QoE).

Meanwhile, a higher ASE means the network can support more users, i.e., with a higher spatial reuse efficiency. In this section, we investigate the interplay of these two metrics in more general multiuser MIMO HetNets. Since the ASE depends on $p_c(k)$, we first derive it in Sect. 5.2.2, and then investigate the trade-off between ASE and p_c in Sect. 5.2.3.

5.2.2 Coverage Analysis

In this subsection, we first derive an exact expression of the coverage probability. Then, an asymptotic expression will be provided, which will expose its key properties in MIMO HetNets.

Since the coverage probability is different if the typical user is associated with different tiers, the overall coverage probability is given by

$$p_c = \sum_{k=1}^{K} A_k p_c(k), \tag{5.44}$$

where $p_c(k)$ is the coverage probability when the user is served by the BS in the k-th tier. Then, we obtain a closed-form expression for the coverage probability $p_c(k)$ based on (5.41) and (5.42), given in the following theorem.

Theorem 5.2 *The coverage probability of the typical user served by the BS in the k-th tier is given by*

$$p_c(k) = A \left\| \mathbf{C}_{M_k}^{-1} \right\|_1, \tag{5.45}$$

where $A = \sum_{j=1}^{K} \lambda_j P_j^\delta B_j^\delta$, and $\mathbf{C}_{M_k}$ is the $M_k \times M_k$ lower triangular Toeplitz matrix, in which

$$\begin{aligned} c_n = & \frac{\delta}{(\delta - n)\Gamma(n+1)} \sum_{j=1}^{K} \lambda_j P_j^\delta B_j^\delta \frac{\Gamma(U_j + n)}{\Gamma(U_j)} \left(\frac{U_k B_k}{U_j B_j} \tau \right)^n \\ & \times {}_2F_1\left(n + U_j, n - \delta; n + 1 - \delta; -\frac{U_k B_k}{U_j B_j} \tau \right). \end{aligned} \tag{5.46}$$

Proof According to (3.56) in Example (3.4), the log-Laplace transform is given by

$$\eta(s) = -2\pi \sum_{j=1}^{K} \lambda_j \int_{l_j(r_0)}^{\infty} \left(1 - \mathbb{E}_{g_j}[\exp(-s g_j v^{-\alpha})]\right) v \mathrm{d}v, \tag{5.47}$$

where $l_j(r_0)$ is the minimal distance between the typical user and the interfering BSs in the j-th tier. As the typical user is associated with the BS in the k-th tier, according to (5.38), it implies that

$$P_k B_k r_k^{-\alpha} \geq P_j B_j r_j^{-\alpha}, \tag{5.48}$$

for $j \in \mathcal{K}$, which is equivalent to

$$r_j \geq \left(\frac{P_j B_j}{P_k B_k}\right)^{\frac{1}{\alpha}} r_k. \tag{5.49}$$

Therefore, similar to (5.14), the log-Laplace transform (5.47) can be written as

$$\begin{aligned}\eta(s) &= -2\pi \sum_{j=1}^{K} \lambda_j \int_{\left(\frac{P_j B_j}{P_k B_k}\right)^{\frac{1}{\alpha}} r_k}^{\infty} \left(1 - \mathbb{E}_{g_j}[\exp(-s g_j v^{-\alpha})]\right) v \mathrm{d}v \\ &= \pi \sum_{j=1}^{K} \lambda_j \left(\frac{P_j B_j}{P_k B_k}\right)^{\delta} r_k^2 \left\{1 - \mathbb{E}_g \left[{}_1F_1\left(-\delta; 1-\delta; -s\left(\frac{P_k B_k}{P_j B_j}\right) r_k^{-\alpha} g\right)\right]\right\}.\end{aligned} \tag{5.50}$$

Therefore, the coefficients of the power series $T(z)$ in (3.94) are given by

$$\begin{aligned}t_n &= \pi r_k^2 \frac{\delta}{\delta - n} \frac{1}{n!} \sum_{j=1}^{K} \lambda_j \left(\frac{P_j B_j}{P_k B_k}\right)^{\delta} \left(\frac{P_k B_k \tau}{P_j B_j \theta}\right)^{n} \\ &\times \left\{\mathbb{1}(n=0) - \mathbb{E}_g\left[g^n {}_1F_1\left(n - \delta; n + 1 - \delta; -\frac{P_k B_k \tau}{P_j B_j \theta} g\right)\right]\right\}.\end{aligned} \tag{5.51}$$

From [26, Lemma 3], the probability density function of r_k is given by

$$f_{r_k}(r) = 2\pi r \left[\sum_{j=1}^{K} \lambda_j \left(\frac{P_j B_j}{P_k B_k}\right)^{\delta}\right] e^{-\pi r^2 \left[\sum_{j=1}^{K} \lambda_j \left(\frac{P_j B_j}{P_k B_k}\right)^{\delta}\right]}. \tag{5.52}$$

According to (3.97), the power series $\bar{P}(z)$ is given by

$$\bar{P}(z) = \mathbb{E}_{r_k}\left[e^{T(z)}\right] = \frac{\sum_{j=1}^{K} \lambda_j P_j^{\delta} B_j^{\delta}}{C(z)} = \frac{A}{C(z)}, \tag{5.53}$$

where the coefficients of $C(z)$ are given by

$$c_n = \frac{\delta}{\delta - n}\frac{1}{n!}\sum_{j=1}^{K}\lambda_j P_j^{\delta} B_j^{\delta}\left(\frac{P_k B_k \tau}{P_j B_j \theta}\right)^n \mathbb{E}_g\left[g^n {}_1F_1\left(n-\delta; n+1-\delta; -\frac{P_k B_k \tau}{P_j B_j \theta} g\right)\right]$$

$$= \frac{\delta}{(\delta - n)\,\Gamma(n+1)}\sum_{j=1}^{K}\lambda_j P_j^{\delta} B_j^{\delta}\frac{\Gamma(U_j+n)}{\Gamma(U_j)}\left(\frac{U_k B_k}{U_j B_j}\tau\right)^n$$

$$\times {}_2F_1\left(n+U_j, n-\delta; n+1-\delta; -\frac{U_k B_k}{U_j B_j}\tau\right), \tag{5.54}$$

where step (5.54) applies Corollary 3.1 with $\theta = \frac{P_k}{U_k}$, $\kappa = U_j$, and $\beta = \frac{P_j}{U_j}$ according to (5.41) and (5.42).

Theorem 5.2 provides a tractable expression that can be easily evaluated numerically. However, it is still in a complicated form, which makes it difficult to directly observe the effects of different system parameters. To overcome this difficulty, we provide an asymptotic result of p_c as the SIR threshold becomes large, i.e., $\tau \to \infty$. Then, we provide some basic properties of p_c, which can help to better understand multiuser MIMO HetNets.

The Taylor expansion of the coefficients c_i in (5.46) gives

$$c_n = \frac{\delta}{(\delta - n)\,\Gamma(n+1)}\sum_{j=1}^{K}\lambda_j P_j^{\delta} B_j^{\delta}\left(\frac{U_k B_k}{U_j B_j}\tau\right)^{\delta}\frac{\Gamma(n+1-\delta)\,\Gamma(U_j+\delta)}{\Gamma(U_j)} + \mathcal{O}\left(\frac{1}{\tau^{U_j}}\right). \tag{5.55}$$

Then, under the assumption that $\tau \to \infty$, the asymptotic expression of the coverage probability is given in the following theorem.

Theorem 5.3 *The asymptotic expression of $p_c(k)$ as $\tau \to \infty$ is given by*

$$p_c(k) \sim A\tau^{-\delta}\mathrm{sinc}(\delta)\frac{(U_k B_k)^{-\delta}\frac{\Gamma(M_k+\delta)}{\Gamma(M_k)}}{\sum_{j=1}^{K}\lambda_j\left(\frac{P_j}{U_j}\right)^{\delta}\frac{\Gamma(U_j+\delta)}{\Gamma(U_j)}}. \tag{5.56}$$

Proof See Appendix.

Based on Theorem 5.3, the coverage probability of the typical user is given by

$$p_c = \sum_{k=1}^{K} A_k p_c(k) \sim \tau^{-\delta}\mathrm{sinc}(\delta)\frac{\sum_{k=1}^{K}\lambda_k\left(\frac{P_k}{U_k}\right)^{\delta}\frac{\Gamma(M_k+\delta)}{\Gamma(M_k)}}{\sum_{j=1}^{K}\lambda_j\left(\frac{P_j}{U_j}\right)^{\delta}\frac{\Gamma(U_j+\delta)}{\Gamma(U_j)}}. \tag{5.57}$$

In Fig. 5.12, we compare the asymptotic result in (5.57) with the exact numerical result in Theorem 5.2, as well as the simulation result. We consider a three-tier HetNet, and the parameters are provided in the caption of the figure. From Fig. 5.12, we find that the numerical result in Theorem 5.2 is exactly the same as the simulation result. Moreover, the asymptotic result in (5.57) provides as an upper bound when

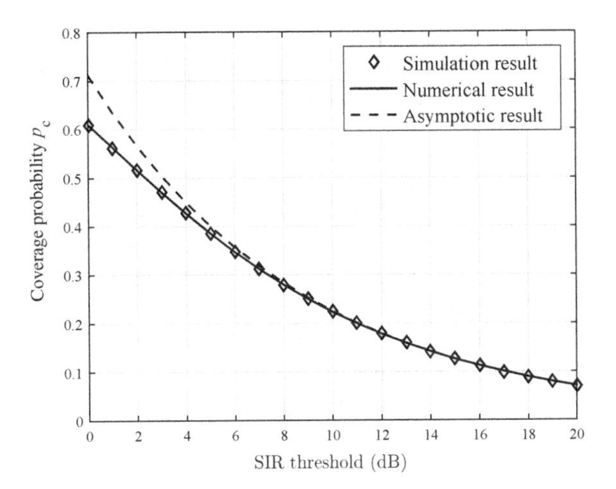

Fig. 5.12 The coverage probability with different SIR threshold, with $\alpha = 4$, $[\lambda_1, \lambda_2, \lambda_3] = [1, 5, 10] \times 10^2$ per km^2, $[M_1, M_2, M_3] = [8, 4, 1]$, $[U_1, U_2, U_3] = [4, 2, 1]$, $[P_1, P_2, P_3] = [6.3, 0.13, 0.05]$W, and $B_k = 1/U_k$.

the SIR threshold τ is not large, but the gap becomes smaller when τ increases. This result confirms the effectiveness of the asymptotic result when the SIR threshold is large. For example, when $\tau \gtrsim 5$ dB in this setting, the asymptotic result is almost the same as the exact result, and above 10 dB, indistinguishable. Note that MIMO techniques will enhance the data rate at relatively high SIRs, so $\tau \gtrsim 5$ dB is practical in MIMO HetNets.

From (5.57), we see a clear relation between the coverage probability and each parameter. It is a perfectly symmetric structure, since both the numerator and denominator are summations of each tier's density $\{\lambda_k\}$,[7] the transmit power to each user $\left\{\frac{P_k}{U_k}\right\}$, and the effects of link array gain $\{M_k\}$ and multiplexing gain $\{U_k\}$, respectively. Interestingly, this expression bears a similar form with the one in [32], which considered SISO HetNets. Based on (5.57), we can easily investigate the effects of different parameters on the coverage probability in the general MIMO HetNets. As a demonstration, we focus on how the BS densities $\{\lambda_k\}$ affect the coverage probability, and present two important properties. For convenience, we rewrite (5.57) in the vector form by defining the column vectors $\boldsymbol{\lambda} = [\lambda_1, \lambda_2, \ldots, \lambda_K]^T$, $\mathbf{c} = [c_1, c_2, \ldots, c_K]^T$ where $c_i = \left(\frac{P_i}{U_i}\right)^{\delta} \frac{\Gamma(M_i+\delta)}{\Gamma(M_i)}$, and $\mathbf{d} = [d_1, d_2, \ldots, d_K]^T$ where $d_i = \left(\frac{P_i}{U_i}\right)^{\delta} \frac{\Gamma(U_i+\delta)}{\Gamma(U_i)}$. Then, (5.57) is equivalent to

$$p_c \sim \tau^{-\delta}\text{sinc}\,(\delta)\,\frac{\mathbf{c}^T\boldsymbol{\lambda}}{\mathbf{d}^T\boldsymbol{\lambda}}. \tag{5.58}$$

Based on (5.58), we obtain the following result.

[7] In this subsection, we use $\{\lambda_k\}$ to denote the set $\{\lambda_k : k \in \mathcal{K}\}$, representing the BS densities of all the tiers, while λ_k is used to denote the BS density of the k-th tier.

Lemma 5.3 *When the SIR threshold τ is large, if the ratio $\frac{c_k}{d_k} = \frac{\Gamma(M_k+\delta)/\Gamma(M_k)}{\Gamma(U_k+\delta)/\Gamma(U_k)}$ is the same for all the tiers, p_c is invariant with λ_k, $\forall k \in \mathcal{K}$; Otherwise, p_c is monotonic with respect to λ_k, i.e., p_c will either increase or decrease as λ_k increases.*

Proof See Appendix.

This result explicitly shows that there is, in general, no SIR invariance property for p_c in MIMO HetNets. Note that previous studies revealed the SIR invariance property in SISO HetNets [26], which means that the SIR distribution is invariant to the BS densities, as long as the mobile connects to the BS providing the strongest received signal power. However, in MIMO HetNets, deploying more BSs of one tier will either increase or decrease the coverage probability. The SIR invariance property that p_c is independent of the BS densities in SISO HetNets is only a special case, where $\frac{c_i}{d_i} = 1$ for all the tiers. Therefore, we should carefully consider how the coverage probability will be affected when densifying the HetNet with multi-antenna BSs.

In this part, we determine the maximum p_c that can be achieved in a given MIMO HetNet. This is equivalent to considering the following optimization problem:

$$\begin{aligned} &\underset{\{\lambda_k\}}{\text{maximize}} && p_c \\ &\text{subject to} && 0 \le \lambda_k \le \lambda_k^{\max}, \forall k \in \mathcal{K}, \end{aligned} \tag{5.59}$$

where λ_k can be regarded as the active BS density of the k-th tier, and $\lambda_k^{\max}$ is the actual BS density. The solution of this optimization problem is provided in Lemma 5.4.

Lemma 5.4 *The maximum p_c with respect to $\{\lambda_k\}$ in MIMO HetNets is given by*

$$p_c^{\max} = p_c(k) \quad \text{for } k = \arg\max_j \frac{\Gamma(M_j+\delta)/\Gamma(M_j)}{\Gamma(U_j+\delta)/\Gamma(U_j)}, \tag{5.60}$$

and the optimal BS density is

$$\lambda_k^{\star} \begin{cases} \in \left(0, \lambda_k^{\max}\right] & k = \arg\max_j \frac{\Gamma(M_j+\delta)/\Gamma(M_j)}{\Gamma(U_j+\delta)/\Gamma(U_j)}, \\ = 0 & k \neq \arg\max_j \frac{\Gamma(M_j+\delta)/\Gamma(M_j)}{\Gamma(U_j+\delta)/\Gamma(U_j)}. \end{cases}$$

Proof See Appendix.

From Lemma 5.4, we see that when the numbers of antennas and the numbers of served users are determined, i.e., when $\{M_k, U_k\}$ are determined, then no matter how to change the BS densities, the coverage probability of the network cannot exceed the value in (5.60). Furthermore, the maximum p_c is achieved by only activating one tier of BSs, which has the largest value of $\frac{\Gamma(M_i+\delta)/\Gamma(M_i)}{\Gamma(U_i+\delta)/\Gamma(U_i)}$.

The two properties above have shown the effect of the BS densities on the coverage probability in a general MIMO HetNet. In the next subsection, we will provide a more comprehensive investigation and jointly analyze the effects of the BS densities on the coverage probability and ASE.

5.2.3 *ASE and Link Reliability Trade-Off*

In this subsection, we investigate the effects of the BS densities on both the link reliability and ASE, and show that there is a trade-off between these two metrics. Moreover, efficient algorithms will be proposed to find the optimal BS densities. To start with, we first consider a special case where all the BSs serve the same number of users. The purpose of investigating such a special case is to reveal the physical insight of the trade-off. Then, we show the trade-off in general MIMO HetNets using the asymptotic results provided in the last section.

We consider the trade-off between the ASE and link reliability in the U-SDMA network, where each BS serves U ($U_k = U \le N_k$ for all $k \in \mathcal{K}$) users. By limiting each BS to serve the same number of users, exact expressions of p_c and ASE are available. Such a network still possesses the key characteristics of the MIMO HetNet. For example, when $U = 1$, the network becomes a TDMA HetNet. For simplicity, we assume $B_k = 1$ for $\forall k \in \mathcal{K}$. Then $p_c(k)$ can be directly obtained from Theorem 5.2, as given in the following corollary.

Corollary 5.1 *The coverage probability of the typical user served by the k-th tier in the U-SDMA HetNet is given by*

$$p_c(k) = \left\| \tilde{\mathbf{C}}_{M_k}^{-1} \right\|_1, \tag{5.61}$$

where the elements $\tilde{c}_n$ *given by*

$$\tilde{c}_n = \frac{\Gamma(U+n)}{\Gamma(U)\Gamma(n+1)} \frac{\delta}{\delta - n} \tau^n {}_2F_1(n-\delta, U+n; n+1-\delta; -\tau). \tag{5.62}$$

Therefore, the coverage probability of the typical user is given by

$$p_c = \sum_{k=1}^{K} A_k p_c(k) = \frac{\sum_{k=1}^{K} \left(\frac{P_k}{U}\right)^{\delta} p_c(k) \lambda_k}{\sum_{k=1}^{K} \left(\frac{P_k}{U}\right)^{\delta} \lambda_k}, \tag{5.63}$$

and its vector form is given by

$$p_c = \frac{\tilde{\mathbf{c}}^T \boldsymbol{\lambda}}{\tilde{\mathbf{d}}^T \boldsymbol{\lambda}}, \tag{5.64}$$

where the k-th elements of $\tilde{\mathbf{c}}$ and $\tilde{\mathbf{d}}$ are $\tilde{c}_k = \left(\frac{P_k}{U}\right)^{\delta} p_c(k)$ and $\tilde{d}_k = \left(\frac{P_k}{U}\right)^{\delta}$, respectively.

From Corollary 5.1, we have the following observations.

Lemma 5.5 *$p_c(k)$ in U-SDMA HetNets has the following properties:*

- *$p_c(k)$ is independent of the BS densities $\{\lambda_k\}$ and the transmit power $\{P_k\}$.*
- *$p_c(i) \geq p_c(j)$ if and only if $M_i \geq M_j$, and the equality holds only if $M_i = M_j$.*

This result means that the tier with more antennas at each BS provides higher link reliability, since the interference suffered by each user has the same distribution in U-SDMA HetNets, while more transmit antennas will provide a higher diversity and array gain. While $p_c(k)$ is independent of both $\{\lambda_k\}$ and $\{P_k\}$, the overall coverage probability p_c in (5.63) depends on $\{\lambda_k, P_k\}$, and p_c can be regarded as a weighted sum of $p_c(k)$ for $k \in \mathcal{K}$. From (5.64), the network coverage probability has the following properties.

Lemma 5.6 *p_c in U-SDMA HetNets has the following properties:*

- *p_c is monotonic with respect to λ_k for $\forall k \in \mathcal{K}$.*
- *The maximum p_c is achieved by only activating one tier of BSs which has the largest number of antennas, i.e.,*

$$p_c^{\max} = p_c(k) \quad \textit{for } k = \arg\max_j M_j. \tag{5.65}$$

Proof Since (5.64) has the same structure with (5.58), Lemmas 5.3 and 5.4 can be applied. Moreover, we have $\frac{\tilde{c}_k}{\tilde{d}_k} = p_c(k)$ in this case, and from Lemma 5.5, we know the maximum $p_c(k)$ is determined by N_k.

By now, it is clear that the coverage probabilities in different tiers ($p_c(k)$) are different, and p_c is an average of each tier's performance. Thus, the densification of different tiers will have different effects on p_c. In particular, increasing the BS density of the tier with a lower $p_c(k)$ will pull down the overall p_c.

On the other hand, the ASE of the K-tier HetNets can be extended from (2.26) and is given by

$$\text{ASE} = \sum_{k=1}^{K} \lambda_k U_k p_c(k) \log_2(1+\tau). \tag{5.66}$$

In the U-SDMA networks, it can be written as

$$\text{ASE} = U \log_2(1+\tau) \sum_{k=1}^{K} \lambda_k p_c(k), \tag{5.67}$$

which shows that increasing the BS density will always increase the ASE.

Note that since $p_c(k)$ is independent of the BS densities, (5.67) shows that activating as many BSs as possible can achieve a higher ASE. Combining the above results, we have the following two conflicting aspects:

- To achieve the maximum ASE, it is optimal to activate all the BSs, but in this case, p_c may be low.

- To achieve the maximum p_c, activating only the tier with the largest number of antennas is optimal, but the ASE will be low.

Thus we need to investigate the trade-off between ASE and link reliability. We are interested in maximizing the achievable ASE given a requirement on p_c, formulated as the following problem:

$$\begin{aligned} \mathscr{P}_o : \underset{\{\lambda_k\}}{\text{maximize}} \quad & \text{ASE} \\ \text{subject to} \quad & p_c \geq \Theta, \\ & 0 \leq \lambda_k \leq \lambda_k^{\max}, \forall k \in \mathscr{K}. \end{aligned} \tag{5.68}$$

Note that in U-SDMA networks, Problem $\mathscr{P}_o$ is a linear programming problem [33], which can be solved efficiently. To get more insights, we derive a more explicit solution. From Lemma 5.6, we know that if $\Theta > \max_k p_c(k)$, there is no feasible solution, which implies that whatever the BS density is, the network cannot achieve such a link reliability requirement. Thus, in the following analysis, we assume $\Theta \leq \max_k p_c(k)$.

Denote $y_k = p_c(k)\lambda_k$, then Problem $\mathscr{P}_0$ is equivalent to

$$\begin{aligned} \mathscr{P}_{U-\text{SDMA}} : \underset{\{y_k\}}{\text{maximize}} \quad & \sum\nolimits_{k=1}^{K} y_k \\ \text{subject to} \quad & \sum\nolimits_{k=1}^{K} b_k y_k \geq 0, \\ & 0 \leq y_k \leq y_k^{\max}, \forall k \in \mathscr{K}, \end{aligned} \tag{5.69}$$

where $b_k = \left(\frac{P_k}{U}\right)^{\delta}\left[1 - \frac{\Theta}{p_c(k)}\right]$. Note that $b_k < 0$ if $p_c(k) < \Theta$. To find the solution, we start from $\left\{y_k^\star = y_k^{\max}\right\}$. If $\sum_{j=1}^{K} b_j y_j^\star \geq 0$, then the optimal solution is to activate all the BSs. On the contrary, if $\sum_{j=1}^{K} b_j y_j^\star < 0$, it means we need to decrease some $y_k^\star$ until $\sum_{j=1}^{K} b_j y_j^\star = 0$. Without loss of generality, assume some of $\{b_k\}$, i.e., $\{b_1, \ldots, b_n\}$ where $n \leq K$, are negative, and $b_1 \leq b_2 \leq b_n < 0$. Then, we need to first decrease $y_1^\star$, as b_1 has the most negative effect. If $y_1^\star = 0$ and $\sum_{j=1}^{K} b_j y_j^\star < 0$, then $y_2^\star$ should be decreased. So on and so forth until $\sum_{j=1}^{K} b_j y_j^\star = 0$. The formal solution is described in Algorithm 1.

From Algorithm 1, we see that the value of b_k is related to the transmit power P_k and the number of antennas N_k. A negative b_k will have a negative effect on p_c. If the BSs in the k-th tier have a smaller N_k and a larger P_k, the value of b_k will be negative and smaller. Therefore, to achieve a high link reliability requirement, the BSs of the tiers with a small b_k should be switched off. Note that one special case is that all the BSs have the same number of antennas, i.e., $N_k = M$ for $\forall k \in \mathscr{K}$. In this special case, $p_c(k)$ is the same for all $k \in \mathscr{K}$. Thus, all the values of $\{b_k\}$ are either greater than 0 or less than 0, and there is no trade-off since p_c becomes a constant. SISO HetNets belong to this special case.

Algorithm 1 Finding the optimal solution of Problem $\mathscr{P}_0$ in U-SDMA HetNets

1: Initialize $y_k^\star \leftarrow y_k^{\max}$ for $k \in \mathscr{K}$ and $n \leftarrow 1$;
2: **while** $\sum_{k=1}^{K} b_k y_k^\star < 0$ **do**
3: $\quad i \leftarrow$ the index of the n-th minimal value among $\{b_k\}$;
4: $\quad y_i^\star \leftarrow 0$;
5: $\quad$ **if** $\sum_{k=1}^{K} b_k y_k^\star \geq 0$ **then**
6: $\quad\quad y_i^\star \leftarrow \frac{1}{-b_i} \sum_{k=1}^{K} b_k y_k^\star$;
7: $\quad\quad$ **break**;
8: $\quad$ **end if**
9: $\quad n \leftarrow n + 1$;
10: **end while**

In the following, we demonstrate how to achieve the maximum ASE given the link reliability requirement in a three-tier HetNet, consisting of micro-BSs, pico-BSs, and femto-BSs, where $[M_1, M_2, M_3] = [4, 2, 2]$, $[P_1, P_2, P_3] = [6.3, 0.13, 0.05]$ Watts [34], and the actual BS densities are $\left[\lambda_1^{\max}, \lambda_2^{\max}, \lambda_3^{\max}\right] = [1, 2, 5] \times 10^2$ per km^2. We consider two cases with $U = 1$ and $U = 2$, respectively.

In Fig. 5.13, we show the trade-off between the maximum ASE and the link reliability requirement Θ. As a benchmark, we also consider the SISO HetNet with the same transmit power $\{P_k\}$ and actual BS densities $\left\{\lambda_k^{\max}\right\}$. From Fig. 5.13, we can find that: (1) Compared with the SISO HetNet, deploying multi-antenna BSs can increase both the ASE and link reliability. For example, in the SISO HetNet, p_c is always 0.56 whatever the BS density is. But when using multi-antenna BSs, the network can achieve better ASE and p_c. (2) Comparing the cases with $U = 1$ and $U = 2$, we see that serving one user at each time slot (i.e., TDMA) can obtain a higher maximum p_c. This is because the channel gains of the interference links

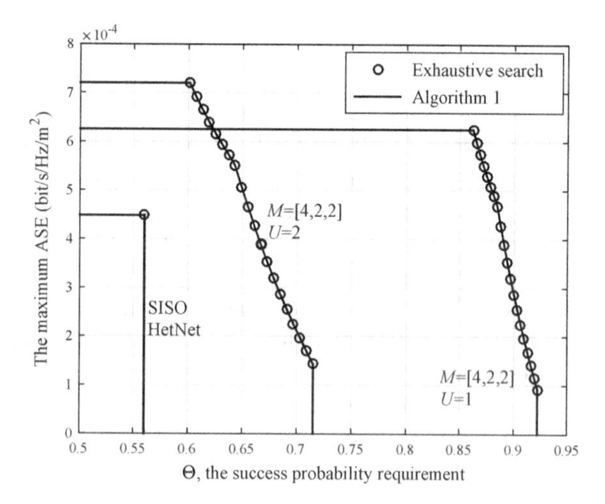

Fig. 5.13 The maximum ASE with different requirements of the coverage probability, with $\alpha = 4$, $\tau = 0$ dB, $[M_1, M_2, M_3] = [4, 2, 2]$, and $[P_1, P_2, P_3] = [6.3, 0.13, 0.05]$ Watts. The actual BS densities are $\left[\lambda_1^{\max}, \lambda_2^{\max}, \lambda_3^{\max}\right] = [1, 2, 5] \times 10^2$ per km^2. © 2016 IEEE. Reprinted, with permission, from [25]

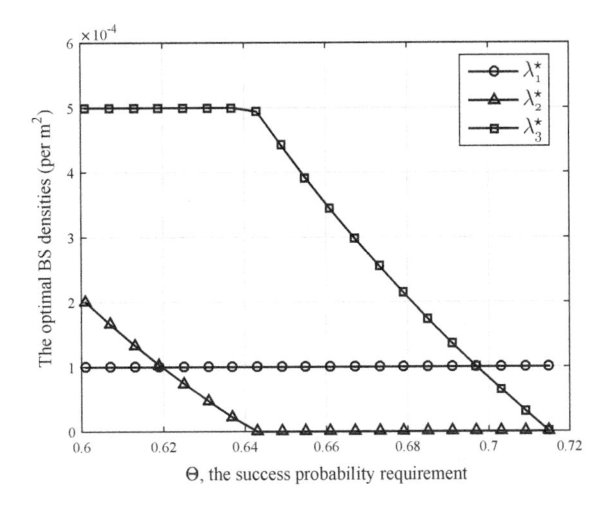

Fig. 5.14 The optimal BS densities with different requirements of the coverage probability, with $\alpha = 4$, $\tau = 0$ dB, $[M_1, M_2, M_3] = [4, 2, 2]$, $U = 2$, and $[P_1, P_2, P_3] = [6.3, 0.13, 0.05]$ Watts. The actual BS densities are $[\lambda_1^{\max}, \lambda_2^{\max}, \lambda_3^{\max}] = [1, 2, 5] \times 10^2$ per km^2. ©2016 IEEE. Reprinted, with permission, from [25]

are small when U is small. However, the maximum ASE is higher when $U = 2$, compared with $U = 1$, since there are more links in a unit area.

The corresponding optimal BS densities of the case with $U = 2$ are provided in Fig. 5.14. When $\Theta > 0.58$, using $\{\lambda_k^{\max}\}$ cannot satisfy the link reliability constraint. From calculation, we find that the second tier has the minimal value of $\left(\frac{P_k}{U}\right)^\delta \left[1 - \frac{\Theta}{p_c(k)}\right]$, which means we need to first decrease the BS density of the second tier, as confirmed in Fig. 5.14. Comparing the third tier with the second tier, we find that $M_2 = M_3$, but $P_2 > P_3$, which means that BSs in the second tier have more negative effects on the link reliability as they will cause higher interference. This is the reason why we need to decrease the BS density of the second tier to improve the link reliability.

Now, we concentrate on the trade-off in a general multiuser MIMO HetNet. Since the analysis becomes extremely difficult with the exact expression of p_c in (5.45), we resort to the asymptotic expression in (5.56). By substituting (5.56) into (5.66), the ASE is given by

$$\text{ASE} \sim \tau^{-\delta} \text{sinc}(\delta) \log_2(1+\tau) \frac{(\mathbf{c}_1^T \boldsymbol{\lambda})(\mathbf{c}_2^T \boldsymbol{\lambda})}{\mathbf{d}^T \boldsymbol{\lambda}}, \tag{5.70}$$

where $\mathbf{c}_1 = [c_{11}, c_{12}, \ldots c_{1K}]^T$, and its element is given as $c_{1i} = P_i^\delta B_i^\delta$, while $\mathbf{c}_2 = [c_{21}, c_{22}, \ldots c_{2K}]^T$, in which $c_{2i} = U_i^{1-\delta} B_i^{-\delta} \frac{\Gamma(M_i+\delta)}{\Gamma(M_i)}$, and $\mathbf{d}$ has the same expression as in (5.58).

From (5.70), we find that the ASE is a quadratic-over-linear function with respect to $\boldsymbol{\lambda}$. It means, in general, we cannot guarantee that increasing the BS density will always increase the ASE, and the effect of the BS density on the ASE becomes complicated. In the following, we consider the same problem $\mathscr{P}_o$ as in (5.68), to find the optimal BS densities that can achieve the maximum ASE. Based on (5.58) and

(5.70), the optimization problem is given as

$$\mathscr{P}_1 : \underset{\boldsymbol{\lambda}}{\text{maximize}} \quad \frac{\left(\mathbf{c}_1^T\boldsymbol{\lambda}\right)\left(\mathbf{c}_2^T\boldsymbol{\lambda}\right)}{\mathbf{d}^T\boldsymbol{\lambda}} \tag{5.71}$$
$$\text{subject to} \quad \tau^{-\delta}\text{sinc}\,(\delta)\,\frac{\mathbf{c}^T\boldsymbol{\lambda}}{\mathbf{d}^T\boldsymbol{\lambda}} \geq \Theta,$$
$$0 \leq \lambda_k \leq \lambda_k^{\max}, k \in \mathscr{K}.$$

Since the maximum p_c is obtained by activating only one tier which has the maximal value of $\frac{\Gamma(M_i+\delta)/\Gamma(M_i)}{\Gamma(U_i+\delta)/\Gamma(U_i)}$ (cf. Lemma 5.4), and $p_c^{\max} \sim \tau^{-\delta}\text{sinc}\,(\delta)\max_i \frac{\Gamma(M_i+\delta)/\Gamma(M_i)}{\Gamma(U_i+\delta)/\Gamma(U_i)}$, we only consider the case when $\Theta \leq p_c^{\max}$.

Problem $\mathscr{P}_1$ is a non-convex problem since the ASE is not a concave function with respect to the BS density. Fortunately, Dinkelbach has proposed an algorithm to solve the nonlinear fractional problems [35]. By defining $N(\boldsymbol{\lambda}) = \left(\mathbf{c}_1^T\boldsymbol{\lambda}\right)\left(\mathbf{c}_2^T\boldsymbol{\lambda}\right)$ and

$$F(t) = \max_{\boldsymbol{\lambda}} \left\{N(\boldsymbol{\lambda}) - t\mathbf{d}^T\boldsymbol{\lambda} \mid \boldsymbol{\lambda} \in \mathscr{S}\right\}, \tag{5.72}$$

where $\mathscr{S} = \left\{\boldsymbol{\lambda} \mid \tau^{-\delta}\text{sinc}\,(\delta)\frac{\mathbf{c}^T\boldsymbol{\lambda}}{\mathbf{d}^T\boldsymbol{\lambda}} \geq \Theta, 0 \leq \lambda_k \leq \lambda_k^{\max}, k \in \mathscr{K}\right\}$, the optimal BS density $\boldsymbol{\lambda}^\star$ can be obtained by finding $t^\star$ such that $F(t^\star) = 0$ [35]. Specifically, by iterating $t^{(i)} = N\left(\boldsymbol{\lambda}^\star\right)/\left(\mathbf{d}^T\boldsymbol{\lambda}^\star\right)$, where $\boldsymbol{\lambda}^\star$ is the optimal solution of the right-hand side of (5.72), $F\left(t^{(i)}\right)$ will converge to 0 and $\boldsymbol{\lambda}^\star$ will be the optimal BS density. However, the optimization problem in (5.72) is still a non-convex problem due to the non-convex function $N(\boldsymbol{\lambda})$. To resolve this difficulty, we resort to the sequential convex programming (SCP) by approximating $N(\boldsymbol{\lambda})$ with the first-order Taylor expansion, given by $N(\boldsymbol{\lambda}) \approx N\left(\boldsymbol{\lambda}^{(n)}\right) + \nabla N\left(\boldsymbol{\lambda}^{(n)}\right)^T\left(\boldsymbol{\lambda} - \boldsymbol{\lambda}^{(n)}\right)$. It is a simple but effective method and has wide applications, e.g., see [36]. Therefore, given the n-th iterative $\boldsymbol{\lambda}^{(n)}$ and the i-th iterative $t^{(i)}$, the convex optimization problem is given by

$$\mathscr{P}_2\left(\boldsymbol{\lambda}^{(n)}, t^{(i)}\right) : \underset{\boldsymbol{\lambda}}{\text{maximize}}\; N\left(\boldsymbol{\lambda}^{(n)}\right) + \nabla N\left(\boldsymbol{\lambda}^{(n)}\right)^T\left(\boldsymbol{\lambda} - \boldsymbol{\lambda}^{(n)}\right) - t^{(i)}\mathbf{d}^T\boldsymbol{\lambda}$$
$$\text{subject to} \quad \left(\tau^{-\delta}\text{sinc}\,(\delta)\,\mathbf{c}^T - \Theta\mathbf{d}^T\right)\boldsymbol{\lambda} \geq 0,$$
$$0 \leq \boldsymbol{\lambda} \leq \boldsymbol{\lambda}^{\max}.$$

Then, the optimal BS density is obtained by Algorithm 2. Note that since SCP can only obtain a local maximum, we randomly generate multiple initial values of $\boldsymbol{\lambda}^{(0)}$ to find a better solution.

Next, we use Algorithm 2 to evaluate the trade-off between the ASE and the link reliability. We consider a three-tier HetNet, where $[M_1, M_2, M_3] = [8, 4, 1]$, $[U_1, U_2, U_3] = [4, 1, 1]$, $B_k = 1/U_k$ for $k = 1, 2, 3$, $[P_1, P_2, P_3] = [6.3, 0.13, 0.05]$ Watts, and the actual BS densities are $\boldsymbol{\lambda}^{\max} = \left[\lambda_1^{\max}, \lambda_2^{\max}, \lambda_3^{\max}\right] = [1, 5, 10] \times 10^2$ per km^2. For Algorithm 2, we generate 20 randomly initial values of $\boldsymbol{\lambda}^{(0)}$, and set $\varepsilon = 10^{-6}$. In Fig. 5.15, we show the trade-off between the maximum ASE and the

Algorithm 2 The Locally Optimal BS Densities in General MIMO HetNets

1: Initialize $\boldsymbol{\lambda}^{(0)} \leftarrow$ random value between $\left[0, \boldsymbol{\lambda}^{\max}\right]$, $n \leftarrow 0$, $i \leftarrow 0$ and assign ε a small value;
2: $t^{(i)} \leftarrow \frac{(\mathbf{c}_1^T \boldsymbol{\lambda}^{(n)})(\mathbf{c}_2^T \boldsymbol{\lambda}^{(n)})}{\mathbf{d}^T \boldsymbol{\lambda}^{(n)}}$;
3: Solve Problem $\mathscr{P}_2\left(\boldsymbol{\lambda}^{(n)}, t^{(i)}\right)$ and obtain the optimal value $\boldsymbol{\lambda}^{\star}$;
4: **if** $\left\|\boldsymbol{\lambda}^{\star} - \boldsymbol{\lambda}^{(n)}\right\| / \left\|\boldsymbol{\lambda}^{(n)}\right\| \geq \varepsilon$ **then**
5: $\quad n \leftarrow n+1$, $\boldsymbol{\lambda}^{(n)} \leftarrow \boldsymbol{\lambda}^{\star}$, and go to Step 3;
6: **else**
7: $\quad$ **if** $\left(\mathbf{c}_1^T \boldsymbol{\lambda}^{\star}\right)\left(\mathbf{c}_2^T \boldsymbol{\lambda}^{\star}\right) - t^{(i)} \mathbf{d}^T \boldsymbol{\lambda}^{\star} \geq \varepsilon$ **then**
8: $\quad\quad i \leftarrow i+1$, and go to Step 2;
9: $\quad$ **else**
10: $\quad\quad$ **return** $\boldsymbol{\lambda}^{\star}$;
11: $\quad$ **end if**
12: **end if**

link reliability requirement Θ, while the corresponding optimal BS densities are shown in Fig. 5.16.

First, we find from both figures that Algorithm 2 can achieve almost the same performance as using the exhaustive search, while Algorithm 2 runs much faster. Second, from Fig. 5.15, we find that in general cases, there exists a trade-off between the ASE and link reliability, and the higher the link reliability requires, the lower the ASE can be achieved. More interestingly, from Fig. 5.16, we find that: (1) Even in general HetNets, the maximal ASE is achieved by activating all the BSs. (2) With the increasing requirement of the link reliability, the BS density will decrease from one tier to another tier. Moreover, only when the BS density decrease to 0, the BS density from another tier will start to decrease, which is the same as U-SDMA HetNets. Thus, we infer that for a given Θ, different tiers have different influences on the network, and there is an ordering of such influences. Recall that in U-SDMA networks, the

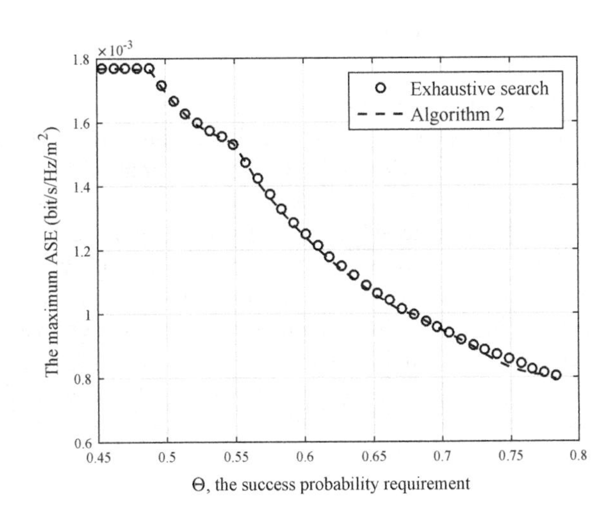

Fig. 5.15 The maximum ASE with different requirements of the coverage probability, with $\alpha = 4$, $\tau = 5$ dB, $[M_1, M_2, M_3] = [8, 4, 1]$, $[U_1, U_2, U_3] = [4, 1, 1]$, $B_k = 1/U_k$ for $k = 1, 2, 3$, and $[P_1, P_2, P_3] = [6.3, 0.13, 0.05]$ Watts. The actual BS densities are $\boldsymbol{\lambda}^{\max} = [1, 5, 10] \times 10^2$ per km^2. ©2016 IEEE. Reprinted, with permission, from [25]

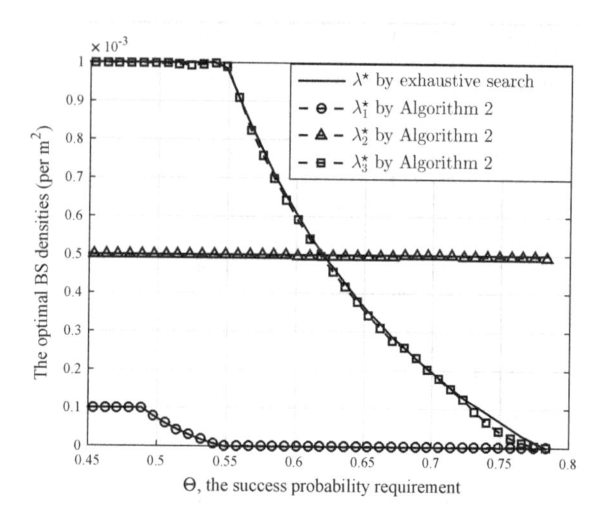

Fig. 5.16 The optimal BS densities with different requirements of the coverage probability, with $\alpha = 4$, $\tau = 5$ dB, $[M_1, M_2, M_3] = [8, 4, 1]$, $[U_1, U_2, U_3] = [4, 1, 1]$, $B_k = 1/U_k$ for $k = 1, 2, 3$, and $[P_1, P_2, P_3] = [6.3, 0.13, 0.05]$ Watts. The actual BS densities are $\lambda^{\max} = [1, 5, 10] \times 10^2$ per km^2. ©2016 IEEE. Reprinted, with permission, from [25]

ordering can be explicitly obtained from $\{b_k\}$, where $b_k = \left(\frac{P_k}{U}\right)^\delta \left[1 - \frac{\Theta}{p_c(k)}\right]$, and the tier with the minimal negative b_k among $k \in \mathcal{K}$ has the most negative effect on the link reliability. However, in the general MIMO HetNets, so far we are unable to derive an exact expression to find such an ordering, since both signal and interference distributions are complicated. But with Algorithm 2, we numerically find the effects of different tiers on the link reliability. For example, from Fig. 5.16, we see that tier 1 (circle points) has the most negative effect on the link reliability. When $\Theta \gtrsim 0.5$, the BSs from tier 1 need to be deactivated to guarantee the link reliability. It is because the interference caused by tier 1 is large due to the high multiplexing gain $U_1 = 4$.

5.2.4 *Summary*

In this section, we provided exact and asymptotic expressions of the coverage probability and the corresponding ASE for multiuser MIMO HetNets, where multi-antenna BSs use SDMA (ZF precoding) to serve multiple users. We focused on the effect of the BS densities and proved that there is no invariance property for the coverage probability in general MIMO HetNets. A unique trade-off between the link reliability and the ASE was revealed. Based on the analysis, we presented effective algorithms to find the optimal BS densities to achieve the maximum ASE while guaranteeing a given link reliability requirement. It was shown that the maximum ASE of the network is achieved by activating all the BSs, while the maximum link reliability is achieved by activating only one tier of BSs. This application example illustrated the effectiveness of the analytical framework in analyzing and optimizing general multi-antenna HetNets.

The link reliability vs. ASE trade-off analyzed in this section provides a new perspective in designing general MIMO HetNets, which is fundamentally different from SISO HetNets. Particularly, both the link reliability and ASE should be considered when evaluating different transmission techniques for HetNets. While we only investigated the effect of the BS densities, it is also interesting to investigate and design other aspects such as load balances and energy efficiency.

Bibliographical Notes

Given the pivotal role of interference management in wireless networks, especially cellular networks, there have been lots of interests in evaluating different interference management techniques in large-scale wireless networks. In studies by Akoum et al. (2013) and Huang et al. (2013), all the BSs in the network are grouped into disjoint clusters, and each BS will avoid intercell interference to users in other cells within the same cluster with interference nulling beamforming. Coverage probabilities were analyzed via stochastic geometry. Joint transmission was investigated by Tanbourgi et al. (2014) and Nigam et al. (2014), where each user is served by several nearby single-antenna BSs under the assumption that the user data is shared between these BSs with high-capacity backhaul links. Zhang et al. (2014) proposed to avoid intercell interference by serving users in different cells with orthogonal channels. Xinga et al. (2013) proposed a similar idea with the fixed-number-based IN, but the analysis was based on coarse approximations. Li et al. (2015) proposed the user-centric interference nulling strategy which was presented in Sect. 5.1.

There have been numerous studies using the Poisson network model to analyze HetNets, but most of them considered SISO HetNets. Readers can refer to the survey paper by ElSawy et al. (2013). One of the most important findings in SISO HetNets is the SIR invariance property, which implies that cell densification does not degrade the link reliability. Thus, the ASE of the network can be increased indefinitely by deploying more BSs. Dhillon et al. (2013) proposed a stochastic ordering approach to analysis MIMO HetNets, but such a method cannot be used for quantitative analysis, since the SINR and SIR distributions were not provided. Gupta et al. (2014) extended it to incorporate load balancing and the achievable rate. Other notable efforts on MIMO HetNets include works limited to two tiers by Chandrasekhar et al. (2009) and Adhikary et al. (2015), and the analysis of the interference distribution by Heath et al. (2013). Li et al. (2016) provided the exact and asymptotic expressions of the coverage probabilities for multiuser MIMO HetNets based on Toeplitz matrices.

Appendix

Proof of Lemma 5.1

Since $\Psi_u^{(1)}$ is an inhomogeneous PPP with density $\lambda_u^{(1)}(y) = \lambda_b p_{\mathrm{a}} \left(e^{-\pi \lambda_b \frac{\|y\|^2}{\mu^2}} - e^{-\pi \lambda_b \|y\|^2} \right)$, the mean number of points in $\mathbb{R}^2$ is given by [37, Sect. 2.4.3]

$$\bar{K} = \int_{\mathbb{R}^2} \lambda_u^{(1)}(y)\,dy. \tag{5.73}$$

Applying the polar-Cartesian transformation, we have $\mathrm{d}y = r\mathrm{d}r\mathrm{d}\theta$, and $\bar{K}$ can be written as

$$\bar{K} = \int_0^{2\pi}\int_0^{\infty} \lambda_b p_{\mathrm{a}} \left(e^{-\pi\lambda_b \frac{r^2}{\mu^2}} - e^{-\pi\lambda_b r^2}\right) r dr d\theta = p_{\mathrm{a}}\left(\mu^2 - 1\right), \tag{5.74}$$

which is a finite value when μ is finite.

Based on the property of the PPP [37], the number of points of $\Psi_u^{(1)}$ in $\mathbb{R}^2$ is Poisson distributed with mean $\bar{K}$, which completes the proof.

Proof of Theorem 5.3

By substituting (5.55) into the power series $C(z)$, we obtain

$$C(z) = \sum_{i=0}^{\infty} c_i z^i \sim \sum_{j=1}^{K} \lambda_j P_j^{\delta} B_j^{\delta} \left(\frac{U_k B_k}{U_j B_j}\tau\right)^{\delta} \frac{\Gamma(U_j+\delta)}{\Gamma(U_j)} \sum_{i=0}^{\infty} \frac{\delta}{\delta - i} \frac{\Gamma(i+1-\delta)}{\Gamma(i+1)} z^i, \tag{5.75}$$

and it can be expressed as

$$C(z) \sim \sum_{j=1}^{K} \lambda_j P_j^{\delta} B_j^{\delta} \left(\frac{U_k B_k}{U_j B_j}\tau\right)^{\delta} \frac{\Gamma(U_j+\delta)}{\Gamma(U_j)} \Gamma(1-\delta)(1-z)^{\delta}. \tag{5.76}$$

Thus, the power series $\bar{P}(z)$ is given by

$$\bar{P}(z) = \frac{A}{C(z)} \sim \frac{A \frac{1}{\Gamma(1-\delta)} (1-z)^{-\delta}}{\sum_{j=1}^{K} \lambda_j P_j^{\delta} B_j^{\delta} \left(\frac{U_k B_k}{U_j B_j}\tau\right)^{\delta} \frac{\Gamma(U_j+\delta)}{\Gamma(U_j)}}. \tag{5.77}$$

Based on the above expression, the coefficient t_n is given by

$$\bar{p}_n = \frac{1}{n!} \bar{P}(z)^{(n)}(z)\,|_{z=0} \sim \frac{A \frac{1}{n!} (\delta)_n \frac{1}{\Gamma(1-\delta)}}{\sum_{j=1}^{K} \lambda_j P_j^{\delta} B_j^{\delta} \left(\frac{U_k B_k}{U_j B_j}\tau\right)^{\delta} \frac{\Gamma(U_j+\delta)}{\Gamma(U_j)}}. \tag{5.78}$$

As $p_{\mathrm{s}}(k) = \sum_{n=0}^{M_k-1} \bar{p}_n$, and the sum $\sum_{n=0}^{M_k-1} \frac{1}{n!} (\delta)_n = \frac{\Gamma(M_k+\delta)}{\Gamma(1+\delta)\Gamma(M_k)}$, the coverage probability is given by

$$p_{\mathrm{s}}(k) \sim \frac{A \frac{\Gamma(M_k+\delta)}{\Gamma(1-\delta)\Gamma(1+\delta)\Gamma(M_k)}}{\sum_{j=1}^{K} \lambda_j P_j^{\delta} B_j^{\delta} \left(\frac{U_k B_k}{U_j B_j}\tau\right)^{\delta} \frac{\Gamma(U_j+\delta)}{\Gamma(U_j)}}. \tag{5.79}$$

Using the equality $\Gamma(1+\delta)\,\Gamma(1-\delta)=\frac{\pi\delta}{\sin(\pi\delta)}=\frac{1}{\mathrm{sinc}(\delta)}$, we obtain (5.56).

Proof of Lemma 5.3

We consider the function $y=\frac{\mathbf{c}^T\boldsymbol{\lambda}}{\mathbf{d}^T\boldsymbol{\lambda}}$, where $\boldsymbol{\lambda}=[\lambda_1,\lambda_2,\ldots\lambda_K]^T$, $\mathbf{c}=[c_1,c_2,\ldots,c_K]$ and $\mathbf{d}=[d_1,d_2,\ldots,d_K]$. The partial derivative of $f(\boldsymbol{\lambda})$ with respect to λ_i is then given by

$$\frac{\partial y}{\partial\lambda_i}=\frac{\sum_{j=1}^{K}d_i d_j\lambda_j\left(\frac{c_i}{d_i}-\frac{c_j}{d_j}\right)}{\left(\mathbf{d}^T\boldsymbol{\lambda}\right)^2}. \tag{5.80}$$

It shows that changing λ_i will not change the sign of $\frac{\partial y}{\partial\lambda_i}$, i.e., y is monotonic with respect to λ_i. Moreover, y is independent of $\boldsymbol{\lambda}$ if $\frac{c_i}{d_i}=\frac{c_j}{d_j}$ for all $i,j\in\{1,2,\ldots,K\}$.

Proof of Lemma 5.4

In this proof, we consider a more general case where the optimization problem is given by

$$\begin{aligned}\mathscr{P}_{w/\,cons}:\ &\underset{\boldsymbol{\lambda}}{\text{maximize}}\quad \frac{\mathbf{c}^T\boldsymbol{\lambda}}{\mathbf{d}^T\boldsymbol{\lambda}}\\ &\text{subject to } \mathbf{A}\boldsymbol{\lambda}\le\mathbf{b},\\ &\qquad\qquad \boldsymbol{\lambda}\ge\mathbf{0},\end{aligned} \tag{5.81}$$

where $\mathbf{A}\boldsymbol{\lambda}\le\mathbf{b}$ represents an arbitrary linear constraint, $\mathbf{A}$ is a $n\times K$ matrix, and $\mathbf{b}$ is a $n\times 1$ vector with positive elements. We will find the optimal solution $\boldsymbol{\lambda}^\star$ in the following derivation.

First, we consider the optimization problem without the constraint $\mathbf{A}\boldsymbol{\lambda}\le\mathbf{b}$, i.e., the problem is given by

$$\begin{aligned}\mathscr{P}_{w/o\,cons}:\ &\underset{\boldsymbol{\lambda}}{\text{maximize}}\quad \frac{\mathbf{c}^T\boldsymbol{\lambda}}{\mathbf{d}^T\boldsymbol{\lambda}}\\ &\text{subject to } \boldsymbol{\lambda}\ge\mathbf{0}.\end{aligned} \tag{5.82}$$

Without loss of generality, we assume $\frac{c_1}{d_1}\ge\frac{c_2}{d_2}\ge\cdots\ge\frac{c_K}{M_k}$, and then from (5.80), we find that to maximize p_s, λ_K should be 0 since $\frac{\partial p_s}{\partial\lambda_K}\le 0$.

Then, repeating the same procedure, we find $\lambda_{K-1}=0$, $\lambda_{K-2}=0$, and $\lambda_2=0$ successively. Finally, the objective function p_s is equal to $\frac{c_1}{d_1}$ with any $\lambda_1>0$, which is the solution of $\mathscr{P}_{w/o\,cons}$.

Second, we consider the optimization Problem $\mathscr{P}_{w/\,cons}$, and we want to prove that the optimal $\boldsymbol{\lambda}^\star$ of Problem $\mathscr{P}_{w/o\,cons}$ is also the optimal solution of problem $\mathscr{P}_{w/\,cons}$. To do so, we need to prove (1) $\boldsymbol{\lambda}^\star$ is feasible for $\mathscr{P}_{w/\,cons}$, and (2) the optimal solution of $\mathscr{P}_{w/\,cons}$ is $\boldsymbol{\lambda}^\star$.

To prove the feasibility, we assume $\boldsymbol{\lambda}_0^\star$ and $\boldsymbol{\lambda}_1^\star$ are optimal solutions of $\mathscr{P}_{w/o\,cons}$, but $\mathbf{A}\boldsymbol{\lambda}_1^\star>\mathbf{b}$. Since $\mathbf{b}>0$, and $\boldsymbol{\lambda}_0^\star=k\boldsymbol{\lambda}_1^\star$ for any $k>0$, we have $\exists k>0$, $\mathbf{A}\boldsymbol{\lambda}_0^\star=k\mathbf{A}\boldsymbol{\lambda}_1^\star\le\mathbf{b}$, i.e., there exists an optimal solution of $\mathscr{P}_{w/o\,cons}$, which is feasible to $\mathscr{P}_{w/\,cons}$.

Finally, since $\mathscr{P}_{w/\,cons}$ has one more constraint than $\mathscr{P}_{w/o\,cons}$, the solution of $\mathscr{P}_{w/\,cons}$ should be the subset of the solution of $\mathscr{P}_{w/o\,cons}$ and the maximum value of $\mathscr{P}_{w/\,cons}$ will no greater than the maximum value of $\mathscr{P}_{w/o\,cons}$. As we have proved that the optimal solution of $\mathscr{P}_{w/o\,cons}$ is feasible in $\mathscr{P}_{w/\,cons}$, we obtain that the maximum value of $\mathscr{P}_{w/\,cons}$ is equal to $\frac{c_1}{d_1}$.

Now we come back to the general MIMO HetNet case. The original problem (5.59) indicates $\mathbf{A} = \mathbf{I}_K$ and $\frac{c_i}{d_i} = \frac{\Gamma(M_i+\delta)/\Gamma(M_i)}{\Gamma(U_i+\delta)/\Gamma(U_i)}$. Therefore, the maximum p_s is obtained by only deploying one tier which has the maximal value of $\frac{c_i}{d_i}$.

References

1. J.G. Andrews, F. Baccelli, R.K. Ganti, A tractable approach to coverage and rate in cellular networks. IEEE Trans. Commun. **59**, 3122–3134 (2011)
2. D. Gesbert, S. Hanly, H. Huang, S.S. Shitz, O. Simeone, W. Yu, Multi-cell MIMO cooperative networks: a new look at interference. IEEE J. Sel. Areas Commun. **28**, 1380–1408 (2010)
3. J. Zhang, R. Chen, J.G. Andrews, A. Ghosh, R.W. Heath Jr., Networked MIMO with clustered linear precoding. IEEE Trans. Wirel. Commun. **8**, 1910–1921 (2009)
4. D. López-Pérez, I. Guvenc, G. de la Roche, M. Kountouris, T.Q.S. Quek, J. Zhang, Enhanced intercell interference coordination challenges in heterogeneous networks. IEEE Wirel. Commun. **18**, 22–30 (2011)
5. G.J. Foschini, K. Karakayali, R.A. Valenzuela, Coordinating multiple antenna cellular networks to achieve enormous spectral efficiency. IEE Proc. Commun. **153**, 548–555 (2006)
6. H. Huh, A.M. Tulino, G. Caire, Network MIMO with linear zero-forcing beamforming: large system analysis, impact of channel estimation, and reduced-complexity scheduling. IEEE Trans. Inf. Theory **58**, 2911–2934 (2012)
7. E. Björnson, E. Jorswieck, Optimal resource allocation in coordinated multi-cell systems. Found. Trends Commun. Inf. Theory **9**, 113–381 (2013)
8. J. Zhang, J.G. Andrews, Adaptive spatial intercell interference cancellation in multicell wireless networks. IEEE J. Sele. Areas Commun. **28**, 1455–1468 (2010)
9. R. Bhagavatula, R.W. Heath Jr., Adaptive bit partitioning for multicell intercell interference nulling with delayed limited feedback. IEEE Trans. Signal Process. **59**, 3824–3836 (2011)
10. S. Akoum, R.W. Heath Jr., Interference coordination: random clustering and adaptive limited feedback. IEEE Trans. Signal Process. **61**, 1822–1834 (2013)
11. K. Huang, J.G. Andrews, An analytical framework for multicell cooperation via stochastic geometry and large deviations. IEEE Trans. Inf. Theory **59**, 2501–2516 (2013)
12. R. Tanbourgi, S. Singh, J.G. Andrews, F.K. Jondral, A tractable model for noncoherent joint-transmission base station cooperation. IEEE Trans. Wirel. Commun. **13**, 4959–4973 (2014)
13. G. Nigam, P. Minero, M. Haenggi, Coordinated multipoint joint transmission in heterogeneous networks. IEEE Trans. Commun. **62**, 4134–4146 (2014)
14. X. Zhang, M. Haenggi, A stochastic geometry analysis of inter-cell interference coordination and intra-cell diversity. IEEE Trans. Wirel. Commun. **13**, 6655–6669 (2014)
15. P. Xia, C. Liu, J.G. Andrews, Downlink coordinated multi-point with overhead modeling in heterogeneous cellular networks. IEEE Trans. Wirel. Commun. **12**, 4025–4037 (2013)
16. C. Li, J. Zhang, M. Haenggi, K.B. Letaief, User-centric intercell interference nulling for downlink small cell networks. IEEE Trans. Commun. **63**, 1419–1431 (2015)
17. N. Jindal, J.G. Andrews, S. Weber, Multi-antenna communication in ad hoc networks: achieving MIMO gains with SIMO transmission. IEEE Trans. Commun. **59**, 529–540 (2011)
18. S.M. Yu, S. Kim, Downlink capacity and base station density in cellular networks, in *IEEE International Symposium on WiOpt* (May 2013), pp. 119–124

19. D.J. Love, R.W. Heath Jr., V.K.N. Lau, D. Gesbert, B.D. Rao, M. Andrews, An overview of limited feedback in wireless communication systems. IEEE J. Sel. Areas Commun. **26**, 1341–1365 (2008)
20. N. Jindal, MIMO broadcast channels with finite-rate feedback. IEEE Trans. Inf. Theory **52**, 5045–5060 (2006)
21. I. Hwang, B. Song, S.S. Soliman, A holistic view on hyper-dense heterogeneous and small cell networks. IEEE Commun. Mag. **51**, 20–27 (2013)
22. A. Ghosh, N. Mangalvedhe, R. Ratasuk, B. Mondal, M. Cudak, E. Visotsky, T.A. Thomas, J.G. Andrews, P. Xia, H.S. Jo, H.S. Dhillon, T.D. Novlan, Heterogeneous cellular networks: from theory to practice. IEEE Commun. Mag. **50**, 54–64 (2012)
23. G. Bartoli, R. Fantacci, K.B. Letaief, D. Marabissi, N. Privitera, M. Pucci, J. Zhang, Beamforming for small cell deployment in LTE-advanced and beyond. IEEE Wirel. Commun. **21**, 50–56 (2014)
24. S. Singh, H.S. Dhillon, J.G. Andrews, Offloading in heterogeneous networks: modeling, analysis, and design insights. IEEE Trans. Wirel. Commun. **12**, 2484–2497 (2013)
25. C. Li, J. Zhang, J.G. Andrews, K.B. Letaief, Success probability and area spectral efficiency in multiuser MIMO HetNets. IEEE Trans. Commun. **64**, 1544–1556 (2016)
26. H.S. Jo, Y.J. Sang, P. Xia, J.G. Andrews, Heterogeneous cellular networks with flexible cell association: a comprehensive downlink SINR analysis. IEEE Trans. Wirel. Commun. **11**, 3484–3495 (2012)
27. A. Ghosh, J. Zhang, J.G. Andrews, R. Muhamed, *Fundamentals of LTE* (Pearson Education, 2010)
28. H.S. Dhillon, M. Kountouris, J.G. Andrews, Downlink MIMO HetNets: modeling, ordering results and performance analysis. IEEE Trans. Wirel. Commun. **12**, 5208–5222 (2013)
29. V. Chandrasekhar, M. Kountouris, J.G. Andrews, Coverage in multi-antenna two-tier networks. IEEE Trans. Wirel. Commun. **8**, 5314–5327 (2009)
30. M. Kountouris, J.G. Andrews, Downlink SDMA with limited feedback in interference-limited wireless networks. IEEE Trans. Wirel. Commun. **11**, 2730–2741 (2012)
31. K. Hosseini, W. Yu, R.S. Adve, Large-scale MIMO versus network MIMO for multicell interference mitigation. IEEE J. Sel. Topics Signal Process. **8**, 930–941 (2014)
32. H.S. Dhillon, R.K. Ganti, F. Baccelli, J.G. Andrews, Modeling and analysis of K-tier downlink heterogeneous cellular networks. IEEE J. Sel. Areas Commun. **30**, 550–560 (2012)
33. S. Boyd, L. Vandenberghe, *Convex Optimization* (Cambridge University Press, Cambridge U.K., 2004)
34. G. Auer, V. Giannini, C. Desset, I. Godor, P. Skillermark, M. Olsson, M.A. Imran, D. Sabella, M.J. Gonzalez, O. Blume, A. Fehske, How much energy is needed to run a wireless network? IEEE Wirel. Commun. **18**, 40–49 (2011)
35. W. Dinkelbach, On nonlinear fractional programming. Manag. Sci. **13**(7), 492–498 (1967)
36. F. Facchinei, S. Sagratella, G. Scutari, Flexible parallel algorithms for big data optimization, in *IEEE International Conference on Acoustics, Speech, and Signal Process. (ICASSP)* (May 2014), pp. 7208–7212
37. M. Haenggi, *Stochastic Geometry for Wireless Networks* (Cambridge University Press, Cambridge, U.K., 2012)

Chapter 6
Summary and Discussion

Abstract This chapter summaries the book and discusses potential extensions. This book introduces analytical methodologies for large-scale multi-antenna wireless networks. The main analytical results presented in the previous chapters are first summarized. Then, extensions of the presented analytical framework are discussed from two aspects. More general network models are discussed in the first part, including more generic channel models, precoding/combining techniques, cell association strategies, and random spatial network models. Extensions to newly emerged application scenarios are then introduced in the second part, including unmanned aerial vehicle systems, physical layer security-aware networks, and vehicular communications systems.

6.1 Summary

This book investigated the performance analysis of multi-antenna wireless networks. Chapters 1 and 2 provided some background and preliminaries for network analysis. Chapter 1 first introduced three application scenarios for which the latest 5G wireless system is designed, i.e., eMBB, mMTC, and URLLC. Furthermore, three principle approaches to achieve eMBB were presented. In particular, ultradense networks, multi-antenna transmission, and mm-wave spectrum are adopted to enhance the capacity of 5G networks in three dimensions, i.e., spatial spectrum reuse, link spectral efficiency, and transmission bandwidth. The first two approaches are identified to be the dominant themes for capacity increase. In other words, the multi-antenna dense network is the typical setting for future network deployment, and hence the analysis of such networks is critical to understand and design 5G networks. Different models for the spatial distribution of wireless networks were then presented, including the grid model, Wyner model, and Poisson distributed model, which are important basis for the network performance analysis.

Chapter 2 mainly presented the fundamentals of network performance analysis via stochastic geometry. Several commonly adopted point processes for modeling the spatial distribution of wireless networks were introduced, including the BPP and PPP. Detailed analysis of single-antenna wireless networks was then presented based

X. Yu et al., *Stochastic Geometry Analysis of Multi-Antenna Wireless Networks*, https://doi.org/10.1007/978-981-13-5880-7_6

on the homogeneous PPP model. It was shown that calculating the Laplace transform of the aggregated interference is a crucial step in the performance analysis, which is, however, no longer the case in multi-antenna network performance analysis.

Chapter 3 is the core of this book, where a unified framework for multi-antenna network performance analysis was presented. In particular, it was shown that the distribution of the SINR value not only depends on the Laplace transform itself but also on its higher order derivatives. By focusing on the log-Laplace transform instead of the Laplace transform, tractable analytical results were derived for the coverage probabilities of different multi-antenna networks, including ad hoc networks and cellular networks. Main innovations of this framework include the recursive relations between the derivatives of the Laplace transform and the matrix representation of the power series. More importantly, system insights, e.g., the impacts of the transmitter density and the antenna size on the coverage probability, were analytically revealed via the framework.

To show the effectiveness of the general analytical framework presented in Chaps. 3–5 applied it to four application examples. Chapter 4 showed the ability of the general framework to yield tractable analysis and design insights for different types of wireless networks. In particular, the effects of the BS density and antenna size on the coverage probability, network ASE, and energy efficiency of small cell multi-antenna networks were characterized, when the user density is taken into consideration. Then, the coverage probability was derived for mm-wave networks. While the propagation characteristics of mm-wave networks are more complicated than conventional networks, the analytical framework still yields tractable results, based on which the impacts of directional antenna arrays were investigated. In Chap. 5, network optimization based on the general framework was carried out. The coverage probability of small cell networks with a novel interference nulling technique was analyzed, based on which the optimal interference nulling range was analytically determined. In addition, the performance of more general MIMO HetNets was analyzed. It was shown that the SIR invariance property observed in single-antenna HetNets no longer holds. In particular, the maximum success probability is achieved by activating only one tier of BSs, while the maximum ASE is achieved by activating all the BSs. This finding revealed a unique trade-off between the ASE and link reliability in multiuser MIMO HetNets. Efficient algorithms were developed for optimizing the optimal BS densities for each tier of the HetNet.

To clearly demonstrate the analytical results derived by the general framework for multi-antenna network performance analysis, Table 6.1 summaries the results in Chaps. 3–5.

6.2 Discussion

In this section, extensions of the analytical methodologies presented in this book are discussed. We will present the extensions from two aspects in the following subsections. The first one discusses more general wireless network models, while the second part presents several new applications.

Table 6.1 Analytical results in Chaps. 3–5

Result	Description
Theorem 3.1	Finite sum representation of the coverage probability
Theorem 3.2	ℓ_1-Toeplitz matrix representation of the coverage probability
Theorem 3.3	Generalized ℓ_1-Toeplitz matrix representation of the coverage probability
Theorem 3.4	Generalized finite sum representation of the coverage probability
Proposition 3.1	SIR coverage probability of cellular networks with arbitrarily distributed interferer's power gain
Corollary 3.1	SIR coverage probability of cellular networks with gamma distributed interferer's power gain
Proposition 3.2	SIR coverage probability of ad hoc networks with arbitrarily distributed interferer's power gain
Corollary 3.2	SIR coverage probability of ad hoc networks with gamma distributed interferer's power gain
Lemma 3.4	Impact of BS density on the coverage probability in cellular networks
Corollary 3.3	Impact of transmitter density on the coverage probability in cellular networks
Proposition 3.4	Impact of antenna size on the coverage probability in cellular networks with arbitrarily distributed interferer's power gain
Corollary 3.4	Impact of antenna size on the coverage probability in cellular networks with gamma distributed interferer's power gain
Proposition 3.5	Impact of antenna size on the coverage probability in ad hoc networks when $\alpha = 4$
Proposition 3.6	Impact of antenna size on the coverage probability in ad hoc networks for general α
Theorem 4.1	SIR coverage probability of the MISO small cell network
Property 4.2	Impact of BS density on the coverage probability in MISO small cell networks
Property 4.3	Impact of antenna size on the coverage probability in MISO small cell networks
Proposition 4.1	Impact of BS density on the energy efficiency in MISO small cell networks
Proposition 4.2	Impact of antenna size on the energy efficiency in MISO small cell networks
Proposition 4.4	SINR coverage probability of mm-wave ad hoc networks
Proposition 4.5	SINR coverage probability of mm-wave cellular networks
Corollary 4.1	Impact of antenna size on the coverage probability in mm-wave ad hoc networks
Corollary 4.3	Impact of antenna size on the coverage probability in mm-wave cellular networks
Theorem 5.1	SIR coverage probability of MISO small cell networks with interference nulling
Proposition 5.1	SIR coverage probability of MISO small cell networks with interference nulling and limited feedback
Theorem 5.2	SIR coverage probability of the MIMO HetNets
Theorem 5.3	Asymptotic SIR coverage probability of the MIMO HetNets
Corollary 5.1	SIR coverage probability of the U-SDMA MIMO HetNets

6.2.1 More General Network Models

The analytical framework in this book for multi-antenna network analysis is based on the network model presented in Sect. 3.1.1. In the following, we shall present some directions of extending this network model to more general cases, by generalizing different parameters in the basic network model.

Propagation and Noise

For the large-scale path loss, the standard power law path loss model (2.17) was assumed in Sect. 3.1.1. In particular, a single path loss exponent α is adopted. Although the standard path loss model has a long history and is the basic setting for most existing wireless network theory, analysis, simulations, and design, it also leads to unrealistic results in some situations [1], e.g., the two-ray model based networks, clustered networks, and mm-wave networks. Specifically, the standard power law path loss model cannot accurately capture the dependence of the path loss exponent α on the link distance. Thus, there have been lots of interests in considering more general path loss models. For example, a single-antenna network analysis based on multi-slope path loss models was investigated in [2]. Further investigation will be needed to test the applicability and tractability of the analytical framework presented in the book to more general path loss models.

On the other hand, for the small-scale fading $\mathbf{H}$, simple distributions are assumed in Sect. 3.1.1, e.g., Rayleigh fading or Nakagami fading. Nevertheless, with the evolution of modern wireless networks, the small-scale fading becomes more complicated. For example, the Saleh–Valenzuela model [3] is adopted to depict mm-wave MIMO channels. Furthermore, the keyhole channel model is used to describe the indirect channel between the transceiver via the intelligent reflecting surface systems [4]. These channel models result in more complicated distributions for both the signal and interferer's power gains, which is the main difficulty in the network performance analysis. How to extend the current methodologies to such cases is an interesting venue for further research.

Precoder and Combiner

In this book, as well as most studies for wireless network analysis, simple precoding and combining strategies are adopted for tractability. More advanced precoding and combining techniques are expected to be incorporated in wireless networks, and should also be considered in the network performance analysis. For example, hybrid precoding was recently proposed for mm-wave and massive MIMO networks [5, 6]. In particular, due to the consideration of cost and power consumption, two precoders, i.e., a digital baseband precoder and an analog radio frequency precoder, are cascaded to compose the precoder at the transmitter side. The design of such hybrid precoders is more challenging, either with the instantaneous channel information or the statistical one. It has been shown in [7] that the performance analysis is difficult even for the link layer with closed-form solutions to the hybrid precoders. It is expected that the network layer performance analysis with more advanced precoders and combiners is highly nontrivial, which is, nevertheless, helpful for analytically revealing the pros and cons of many newly proposed MIMO transmission techniques.

Cell Association

As shown in Sect. 3.1.1, the receive power of the desired information signal is critically determined by the distance r_0 from the typical receiver to its associated transmitter. In most networks discussed in this book, the nearest-BS association and bipolar association are assumed in cellular and ad hoc networks, respectively. However, with the evolution of the wireless network, the cell association strategy shall no longer only depend on the distance, especially in HetNets. More factors should be taken into consideration for the cell association design, e.g., load balancing [8], interference coordination [9, 10], or some utility functions with specific network performance requirements [11, 12]. It is intriguing to incorporate these novel cell association strategies in the analysis of multi-antenna wireless networks. The main technical challenge for yielding a tractable analytical result is to derive the explicit distribution for the random distance r_0, and some tenable approximations may be needed.

Spatial Model of Interferers

In Sect. 3.1.1, we assumed that the interfering transmitters form a conditional homogeneous PPP. While later in Chaps. 4 and 5, we showed that inhomogeneous PPPs are also compatible with our general framework. However, PPPs are not enough for capturing the characteristics of wireless networks in some occasions. For example, the Poisson cluster process has been adopted to model the HetNets with *correlated* user and BS locations. Furthermore, the Poisson hole process (PHP) introduced in Sect. 2.1.3 is usually used in cognitive radio networks [13, 14], where the primary and secondary users are modeled by the two PPPs in the PHP. More recently, a cox process driven by the Poisson line process has been proposed to model the vehicular-to-vehicular network [15]. Nevertheless, most of these studies focus on single-antenna networks. Therefore, it is valuable to consider extending the analytical framework in this book to more general network models by employing different point processes for the spatial distributions of transceivers.

Performance Analysis of Uplink Transmission

In this book, we mainly focus on the performance analysis of downlink transmission. On the other hand, the performance of uplink transmission is also of great value to characterize for providing a complete understanding of dense wireless networks. However, the large-scale network performance analysis of uplink transmission differs from the downlink case and is highly challenging, even with single-antenna transceivers. In order to make the analysis tractable, several approximations were made in different aspects in early work on uplink analysis [16], e.g., the simple choice of the BS of interest, and the independence assumption on the distances between the users and their serving BSs.

Specifically, the interference in the uplink is inherently more complicated, and power control is the main factor that makes the uplink analysis more difficult. Most stochastic geometry-based studies of the uplink transmission incorporate power control (or fractional power control) to invert the large-scale path loss. One commonly used power control method is truncated channel inversion, where the transmitters compensate for the path loss to keep the average received signal power equal to a constant. In [17], the authors have shown that a tractable paradigm can be derived when each transmitter employs a truncated channel inversion power control policy with a cutoff threshold. In addition, the meta-distribution of the SIR has been derived in [18] based on a considerably more accurate model in the uplink, which also considers downlink power control. Note that these works focused on single-antenna networks with the SINR expression similar to that in [19]. This means our proposed framework has the potential to tackle the uplink analysis of multi-antenna networks where the truncated channel inversion power control is utilized, which is a promising future research direction.

6.2.2 Newly Emerged Application Scenarios

In Chaps. 4 and 5, we applied the analytical framework to MISO small cell networks, mm-wave ad hoc and cellular networks, and U-SDMA MIMO HetNets. Recently, there have been many newly emerged application scenarios resulted from various user and system requirements. In the following, we shall introduce three applications to which the analytical framework can be further applied.

Unmanned Aerial Vehicle (UAV) Systems

Thanks to their flexibility, easy deployment and relatively small operating expenses, we are witnessing an explosion of commercial UAVs, also known as drones. These small aerial vehicles have become part of daily operations across industries, enabling a full continuum of civilian applications ranging from rapid medical supply delivery, precision agriculture, public safety, to search-and-rescue.

Some early attempts of UAV communications systems tried to leverage the mobility of UAVs to enhance the performance of the current cellular networks. In particular, UAVs act as aerial BSs to overcome the limitations of the current system [20–22]. For example, UAVs are adopted to fill the coverage holes in existing cellular networks. Their mobility can also provide flexible and dynamic coverage to terrestrial users [23]. To analyze such networks consisting of both terrestrial and aerial BSs, how to model the spatial distribution of these BSs is a key problem. A 3D random spatial model is required to characterize the aerial and terrestrial BSs. In this way, the distribution of the aggregated interference from interfering BSs both on the ground and in the sky is more complicated.

On the other hand, for the effective control, monitoring, and navigation of UAVs, connectivity with the ground control station via wireless communications is essential, which helps to exchange information including flight status, control command, and sensing messages [24]. In this case, the UAVs act as aerial users rather than BSs. While the interfering BSs are all on the ground, the performance analysis of large-scale cellular-assisted UAV networks is still challenging. Since the BS antennas are typically down tilted toward ground users, directional beams synthesized by multiple antennas are often used at the UAVs to select the serving BSs with large enough power gains. In addition, a probabilistic path loss model is widely adopted, for which LOS or NLOS propagations occur with different probabilities, depending on such effects as the distance and heights of transceivers, and the density and height of buildings [25].

As a short summary, when UAVs either act as aerial BSs or users, compared with the 2D network considered in this book, the 3D link distances, as well as other effects such as BS antenna tilting and directional antennas, bring additional difficulties. It is valuable to test if the analytical framework in this book can be extended to 3D networks. In particular, such an analysis will help to evaluate the effectiveness of different multi-antenna techniques for UAV communications.

Physical Layer Security-Aware Networks

Next-generation wireless networks will be characterized by dense deployment with small cells, together with massive connected devices, such as wearable devices and smart-house devices. Such highly complex networks will put information security as one primary concern. Physical layer security mechanisms [26, 27] have drawn ever-increasing attention, and have witnessed significant growth [28–30].

For the performance analysis of physical layer security-aware networks, there are two important performance metrics. The first one is the *connection outage probability*, which is defined by the probability that the SINR of the typical receiver is less than a threshold. This outage probability is similar to that in conventional network analysis. The other performance metric is the *secrecy outage probability*. The message is not perfectly secure against the eavesdropper if the SINR at the eavesdropper is larger than a threshold. This outage probability is unique for the physical layer security-aware networks and therefore needs careful investigation.

While interference is conventionally considered deleterious for communications, it was revealed in [31] that the interference can be beneficial for network secrecy because interference makes the eavesdroppers more difficult to intercept transmissions. To enhance the secrecy performance, cooperative jamming [32, 33] and artificial noise assisted [34–36] methods were introduced, both aiming to generate additional interference to the eavesdroppers. To fully exploit the potential of interference in secrecy protection, it is necessary to generate strong interference at the eavesdroppers while having mild effect at the legitimate receivers. Fortunately, the multiple antennas can provide benefits in this aspect and therefore play a pivotal role in physical layer security-aware networks. Advanced techniques can be developed for both generating jamming to the eavesdroppers and enhancing the desired signal power to the legitimate receiver. The analytical framework in this book would be a good candidate for designing and analyzing the performance of such techniques in physical layer security-aware networks.

Vehicular Communications Systems

Vehicular communications, which refers to vehicle-to-vehicle (V2V) and vehicle-to-infrastructure (V2I) communications, has enabled the vehicular nodes to transmit and receive information with each other, as well as with roadside units, to improve the road safety and transport efficiency [37].

In Sect. 3.1.1, PPPs were adopted to model the spatial distributions of the transceivers in wireless networks. In contrast, for the performance analysis of vehicular communications systems, the network model is quite different from conventional networks. A cluster-based model was recently introduced to model the heterogeneous vehicular networks in [38], based on which the performance analysis of uplink transmission was carried out. Furthermore, a Poisson line process was adopted in [15, 39] to model the road network and, conditionally on the lines, and PPPs are used to model the vehicles on the roads. By modeling the vehicular networks in this way, two different Palm distributions with respect to the PPPs need to be investigated, which brings formidable challenges.

In addition to the spatial random model, another key differentiating characteristic in vehicular communications systems is the node mobility. In particular, due to the mobility nature of vehicles and the related high vehicular speed, the topology of vehicular wireless networks becomes highly dynamic [40]. In this case, cooperative transmissions are recommended as a promising solution for vehicles in 5G cooperative small cell networks. This typical setting makes the distribution of the desired signal power gain more complicated, which is assumed to be the (generalized) gamma distribution is this book. Therefore, it is valuable to extend the analytical framework in this book to incorporate cooperative transmissions.

Similar to UAV communications systems, reliable connectivity with the current cellular network shall help the vehicular communications systems to improve their scalability, enhance the high mobility support, and meet the latency requirements [41]. Hence, it is also of great importance to analytically depict the performance of cellular-assisted vehicular systems. In this case, incorporating both the point/line processes for cellular and vehicular networks in the analytical framework is a crucial task.

References

1. H. Inaltekin, M. Chiang, H.V. Poor, S.B. Wicker, On unbounded path-loss models: effects of singularity on wireless network performance. IEEE J. Sel. Areas Commun. **27**, 1078–1092 (2009)
2. X. Zhang, J.G. Andrews, Downlink cellular network analysis with multi-slope path loss models. IEEE Trans. Commun. **63**, 1881–1894 (2015)
3. T.S. Rappaport, R.W. Heath Jr., R.C. Daniels, J.N. Murdock, *Millimeter Wave Wireless Communications* (Pearson Education, 2014)
4. J.D. Griffin, G.D. Durgin, Complete link budgets for backscatter-radio and RFID systems. IEEE Antennas Propag. Mag. **51**, 11–25 (2009)

5. O.E. Ayach, S. Rajagopal, S. Abu-Surra, Z. Pi, R.W. Heath Jr., Spatially sparse precoding in millimeter wave MIMO systems. IEEE Trans. Wirel. Commun. **13**, 1499–1513 (2014)
6. X. Yu, J. Shen, J. Zhang, K.B. Letaief, Alternating minimization algorithms for hybrid precoding in millimeter wave MIMO systems. IEEE J. Sel. Topics Signal Process. **10**, 485–500 (2016)
7. L. Liang, W. Xu, X. Dong, Low-complexity hybrid precoding in massive multiuser mimo systems. IEEE Wirel. Commun. Lett. **3**, 653–656 (2014)
8. H.S. Jo, Y.J. Sang, P. Xia, J.G. Andrews, Heterogeneous cellular networks with flexible cell association: a comprehensive downlink SINR analysis. IEEE Trans. Wirel. Commun. **11**, 3484–3495 (2012)
9. R. Madan, J. Borran, A. Sampath, N. Bhushan, A. Khandekar, T. Ji, Cell association and interference coordination in heterogeneous LTE-A cellular networks. IEEE J. Sel. Areas Commun. **28**, 1479–1489 (2010)
10. K. Son, S. Chong, G.D. Veciana, Dynamic association for load balancing and interference avoidance in multi-cell networks. IEEE Trans. Wirel. Commun. **8**, 3566–3576 (2009)
11. D. Bethanabhotla, O.Y. Bursalioglu, H.C. Papadopoulos, G. Caire, Optimal user-cell association for massive MIMO wireless networks. IEEE Trans. Wirel. Commun. **15**, 1835–1850 (2016)
12. H. Boostanimehr, V.K. Bhargava, Unified and distributed QoS-driven cell association algorithms in heterogeneous networks. IEEE Trans. Wirel. Commun. **14**, 1650–1662 (2015)
13. C. Lee, M. Haenggi, Interference and outage in poisson cognitive networks. IEEE Trans. Wirel. Commun. **11**, 1392–1401 (2012)
14. Z. Yazdanshenasan, H.S. Dhillon, M. Afshang, P.H.J. Chong, Poisson hole process: theory and applications to wireless networks. IEEE Trans. Wirel. Commun. **15**, 7531–7546 (2016)
15. V.V. Chetlur, H.S. Dhillon, Coverage analysis of a vehicular network modeled as cox process driven by Poisson line process. IEEE Trans. Wirel. Commun. **17**, 4401–4416 (2018)
16. T.D. Novlan, H.S. Dhillon, J.G. Andrews, Analytical modeling of uplink cellular networks. IEEE Trans. Wirel. Commun. **12**, 2669–2679 (2013)
17. H. ElSawy, E. Hossain, On stochastic geometry modeling of cellular uplink transmission with truncated channel inversion power control. IEEE Trans. Wirel. Commun. **13**, 4454–4469 (2014)
18. Y. Wang, M. Haenggi, Z. Tan, The meta distribution of the SIR for cellular networks with power control. IEEE Trans. Commun. **66**, 1745–1757 (2018)
19. J.G. Andrews, F. Baccelli, R.K. Ganti, A tractable approach to coverage and rate in cellular networks. IEEE Trans. Commun. **59**, 3122–3134 (2011)
20. Q. Wu, Y. Zeng, R. Zhang, Joint trajectory and communication design for multi-UAV enabled wireless networks. IEEE Trans. Wirel. Commun. **17**, 2109–2121 (2018)
21. M. Mozaffari, W. Saad, M. Bennis, M. Debbah, Mobile unmanned aerial vehicles (UAVs) for energy-efficient internet of things communications. IEEE Trans. Wirel. Commun. **16**, 7574–7589 (2017)
22. O. Esrafilian, R. Gangula, D. Gesbert, Learning to communicate in UAV-aided wireless networks: map-based approaches. IEEE Internet Things J. (to appear)
23. M. Alzenad, A. El-Keyi, F. Lagum, H. Yanikomeroglu, 3-d placement of an unmanned aerial vehicle base station (UAV-BS) for energy-efficient maximal coverage. IEEE Wirel. Commun. Lett. **6**, 434–437 (2017)
24. Y. Zeng, J. Lyu, R. Zhang, Cellular-connected UAV: Potential, challenges and promising technologies. IEEE Wirel. Commun. (to appear)
25. A. Al-Hourani, S. Kandeepan, A. Jamalipour, Modeling air-to-ground path loss for low altitude platforms in urban environments, in *IEEE Global Communications Conference (GLOBECOM)* (Dec 2014), pp. 2898–2904
26. A. Wyner, The wire-tap channel. Bell Syst. Tech. J. **54**, 1355–1387 (1975)
27. I. Csiszár, J. Körner, Broadcast channels with confidential messages. IEEE Trans. Inf. Theory **24**, 339–348 (1978)
28. A. Mukherjee, S.A.A. Fakoorian, J. Huang, A.L. Swindlehurst, Principles of physical layer security in multiuser wireless networks: a survey. IEEE Commun. Surv. Tuts. **16**, 1550–1573 (2014) (3rd Quart.)

29. Y. Shiu, S.Y. Chang, H. Wu, S.C.-H. Huang, H. Chen, Physical layer security in wireless networks: a tutorial. IEEE Wirel. Commun. **18**, 66–74 (2011)
30. H. Wang, T. Zheng, J. Yuan, D. Towsley, M.H. Lee, Physical layer security in heterogeneous cellular networks. IEEE Trans. Commun. **64**, 1204–1219 (2016)
31. A. Rabbachin, A. Conti, M.Z. Win, Wireless network intrinsic secrecy. IEEE/ACM Trans. Netw. **23**, 56–69 (2015)
32. E. Tekin, A. Yener, The general Gaussian multiple-access and two-way wiretap channels: achievable rates and cooperative jamming. IEEE Trans Inf. Theory **54**, 2735–2751 (2008)
33. X. Zhou, M. Tao, R.A. Kennedy, Cooperative jamming for secrecy in decentralized wireless networks, in *Proceedings of the IEEE International Conference on Communications (ICC)*, (Ottawa, Canada, 2012), pp. 2339–2344
34. S. Goel, R. Negi, Guaranteeing secrecy using artificial noise. IEEE Trans. Wirel. Commun. **7**, 2180–2189 (2008)
35. W. Shi, J.A. Ritcey, Distributed jamming for secure communication in Poisson fields of legitimate nodes and eavesdroppers, in *Proceedings of the Asilomar Conference on Signals, Systems, and Computers*, (Pacific Grove, CA, 2012), pp. 1881–1885
36. X. Zhang, X. Zhou, M.R. McKay, Enhancing secrecy with multi-antenna transmission in wireless ad hoc networks. IEEE Trans. Inf. Forensics Secur. **8**, 1802–1814 (2013)
37. S. Biswas, R. Tatchikou, F. Dion, Vehicle-to-vehicle wireless communication protocols for enhancing highway traffic safety. IEEE Commun. Mag. **44**, 74–82 (2006)
38. Q. Zheng, K. Zheng, L. Sun, V.C.M. Leung, Dynamic performance analysis of uplink transmission in cluster-based heterogeneous vehicular networks. IEEE Trans. Veh. Tech. **64**, 5584–5595 (2015)
39. C. Choi, F. Baccelli, An analytical framework for coverage in cellular networks leveraging vehicles. IEEE Trans. Commun. **66**, 4950–4964 (2018)
40. Y. Zhu, Y. Bao, B. Li, On maximizing delay-constrained coverage of urban vehicular networks. IEEE J. Sel. Areas Commun. **30**, 804–817 (2012)
41. S. Schwarz, T. Philosof, M. Rupp, Signal processing challenges in cellular-assisted vehicular communications: Efforts and developments within 3GPP LTE and beyond. IEEE Signal Process. Mag. **34**, 47–59 (2017)

Printed in the United States
By Bookmasters